Contents

List of plates *vi*
List of figures *viii*
List of tables *x*
List of boxes *xi*
Acknowledgements *xii*
List of abbreviations *xiv*

1 Introduction: why disaster and development? 1
2 Viewing disasters from perspectives of development 46
3 How do disasters influence development? 88
4 Physical and mental health in disaster and development 123
5 Learning and planning in disaster management 176
6 Disaster early warning and risk management 205
7 Disaster mitigation, response and recovery 227
8 Conclusion 250

Bibliography 263
Index 279

Plates

1.1	The cyclone defence wall at Chakoria, Bangladesh	9
1.2	Landslide remediation in Nepal	9
1.3	Remains of a residence in Mozambique following coastal erosion and storms	10
1.4	A bombed bridge in Mozambique being used to dry clothes	10
1.5	Tsunami damage of a community centre in Sri Lanka	22
1.6	High-rise and low-rise hazards and vulnerabilities in Dhaka, Bangladesh	29
1.7	Save the Children, an international NGO that helps children in development and in disasters	38
2.1	Students learning about forest management in Scotland	54
2.2	A livelihood from woodland clearance and charcoal production in Africa	54
2.3	The National Memorial of Bangladesh. Several million victims of conflict and famine in the 1970s are buried at this site	66
2.4	The remains of a Soviet tank in Africa	67
2.5	The recently expanded cityscape of Kathmandu, Nepal. It is in an earthquake zone	83
2.6	One small corner of urban development at Beijing, China, will bring many from the rural areas to live in these apartments	84
3.1	An Internally Displaced Persons (IDP) camp in Eastern Sri Lanka	93

Disaster and Development

Andrew E. Collins

LONDON AND NEW YORK

First published 2009
by Routledge
2 Park Square, Milton Park, Abingdon, Oxon, OX14 4RN

Simultaneously published in the USA and Canada
by Routledge
270 Madison Avenue, New York, NY 10016

Routledge is an imprint of the Taylor & Francis Group, an informa business

Typeset in Times New Roman by Keyword Group Ltd
Printed and bound in Great Britain by T. J. International Ltd., Padstow, Cornwall

British Library Cataloguing in Publication Data
A catalogue record for this book is available from the British Library

Library of Congress Cataloguing in Publication Data
Collins, Andrew E.
Disaster and development/Andrew E. Collins.
p.cm.
Includes bibliographical references and index.
1. Economic development. 2. Natural disasters. I. Title.
HD45.C62 2009
338.9–dc22
2008044851

ISBN 13: 978-0-415-42667-1 (hbk)
ISBN 13: 978-0-415-42668-8 (pbk)
ISBN 13: 978-0-203-87923-8 (ebk)

ISBN 10: 0-415-42667-7 (hbk)
ISBN 10: 0-415-42668-5 (pbk)
ISBN 10: 0-203-87923-6 (ebk)

Disaster and Development

Development to a large extent determines the way in which hazards impact on people. Meanwhile the occurrence of disasters alters the scope of development. Whilst a notion of the association of disaster and development is as old as development studies itself, recent decades have produced an intensifying demand for a fuller understanding. Evidence of disaster and development progressing together has attracted increased institutional attention. This includes recognition, through global accords, of a need for disaster reduction in achieving Millennium Development Goals, and of sustainable development as central to disaster reduction. However, varied interpretations of this linkage, and accessible options for future human wellbeing, remain unconsolidated for most of humanity.

This engaging and accessible text illuminates the complexity of the relationship between disaster and development; it opens with an assessment of the scope of contemporary disaster and development studies, highlighting the rationale for looking at the two issues as part of the same topic. The second and third chapters detail development perspectives of disaster, and the influence of disaster on development. The fourth chapter exemplifies how human health is both a cause and consequence of disaster and development and the following chapter illustrates some of the learning and planning processes in disaster and development oriented practice. Early warning, risk management, mitigation, response and recovery actions provide the focus for the sixth and seventh chapters. The final chapter indicates some of the likely future contribution and challenges of combined disaster and development approaches. With an emphasis on putting people at the centre of disaster and development, the book avoids confronting readers with 'no hope' representations, instead highlighting disaster reduction opportunities.

This book is an essential introduction for students from multiple disciplines, whose subject area may variously engage with contemporary crises, and for many other people interested in finding about what is really meant by disaster reduction. They include students and practitioners of development, environment, sociology, economics, public health, anthropology, and emergency planning, amongst others. It provides an entry point to a critical, yet diverse topic, backed up by student-friendly features, such as boxed case studies from the geographical areas of America to Africa and parts of Europe to parts of the East, summaries, discussion questions, suggested further reading and web site information.

Dr Andrew Collins is Reader at Northumbria University. He has engaged in disaster and development logistics and research internationally since 1986, led establishment of a postgraduate programme in disaster management and sustainable development and the Disaster and Development Centre (DDC) launched 2000 and 2004 respectively.

Routledge Perspectives on Development

Series Editor: Professor Tony Binns, *University of Otago*

The *Perspectives on Development* series will provide an invaluable, up to date and refreshing approach to key development issues for academics and students working in the field of development, in disciplines such as anthropology, economics, geography, international relations, politics and sociology. The series will also be of particular interest to those working in interdisciplinary fields, such as area studies (African, Asian and Latin American Studies), development studies, rural and urban studies, travel and tourism.

If you would like to submit a book proposal for the series, please contact Tony Binns on j.a.binns@geography.otago.ac.nz

Published:

David W. Drakakis-Smith
Third World Cities, 2nd edition

Kenneth Lynch
Rural-Urban Interactions in the Developing World

Nicola Ansell
Children, Youth & Development

Katie Willis
Theories and Practices of Development

Jennifer A. Elliott
An Introduction to Sustainable Development, 3rd edition

Chris Barrow
Environmental Management & Development

Janet Henshall Momsen
Gender and Development

Richard Sharpley and David J. Telfer
Tourism and Development

Andrew McGregor
Southeast Asian Development

Cheryl McEwan
Postcolonialism and Development

Andrew Williams & Roger MacGinty
Conflict and Development

Andrew Collins
Disaster and Development

Forthcoming:

Jo Beall and Sean Fox
Cities and Development

W.T.S. Gould
Population and Development

David Lewis and Nazneen Kanji
Non-Governmental Organisations and Development

Janet Henshall Momsen
Gender and Development Second Edition

Clive Agnew and Philip Woodhouse
Water Resources and Development

David Hudson
Global Finance and Development

Michael Tribe, Frederick Nixon and Andrew Sumner
Economics and Development Studies

Tony Binns and Alan Dixon
Africa: Diversity and Development

Tony Binns, Christo Fabricius and Etienne Nel
Local Knowledge, Environment and Development

Andrea Cornwall
Participation and Development

Hazel Barrett
Health and Development

3.2 Cyclone shelter in coastal Bangladesh that was donated by the Saudi Arabian Government. This is used as a school, mosque and community centre when there are no cyclones. Its combined functions go some way to exemplifying what might be meant by infrastructural and developmental resilience 105
3.3 Women on the move in Western Darfur, Sudan 112
4.1 Community volunteers cleaning the neighbourhood in Beira, Mozambique 137
4.2 An open well at Beira, Mozambique 141
4.3 An arsenic contaminated bore hole in Bangladesh. When a bore hole is contaminated, people are told to revert to surface water, which needs to be treated 142
4.4 The cholera ward at Beira, Mozambique 144
4.5 Poster to feed back information on health risks to a community in Bangladesh 168
4.6 Sharing a risk analysis with part of a community at Beira, Mozambique 168
5.1 A participatory appraisal in East Timor 184
5.2 A child takes part in an activity for orphans and vulnerable children (OVC) that allows expression of likes, hopes and aspirations 186
6.1 Inside one of the many communication centres of the Cyclone Preparedness Programme in Bangladesh 207
6.2 Monitoring ecological early warning indicators, Bangladesh. Changes in algae in this pond may be associated with a heightened likelihood of cholera 210
6.3 Monitoring food security through a household-based interview in Mozambique 211
7.1 Food aid being delivered along the Zambezi river 246
7.2 House reconstruction in post-tsunami Sri Lanka 247

Figures

1.1	Disaster and development studies	3
1.2	The *World Disasters Report* is an annual release of the International Federation of Red Cross and Red Crescent Societies	12
1.3	Total number of disasters reported 1900–2006	13
1.4	Multilevelled development analysis	23
1.5	Disaster management cycle and development	27
2.1	Rational influences on individual and group environmental behaviour	52
2.2	Poster marking the Third Congress of the ruling FRELIMO party in Mozambique, 1977. In English it reads 'Win the Battle of Production' and then 'Long live the 3rd congress. Long live FRELIMO, the party of the Mozambique people'	68
2.3	Generalised hypothesis on the link between poverty and environmental degradation	71
2.4	An integrated poverty and environment view of humanitarian disasters	73
2.5	Idealised reversal of poverty, environment and disasters processes	75
2.6	Household sustainable development for poverty reduction	78
3.1	Security and livelihoods: (a) increasing human security, (b) diminishing human security, where H = human capital, S = social capital, N = natural capital, Pol = political capital, F = financial capital and P = physical capital	100

4.1 Health hazards, vulnerability and care 131
4.2 The health ecology approach to infectious disease risk reduction, for health security at global, community and individual levels 136
5.1 Learning cycle based on project planning 178
5.2 Communications within a community-based risk and resilience programme 189
5.3 Indicative (idealised) stakeholder interactions relative to an empowered community-based organisation 189
5.4 Summary logical framework analysis for a project on risk and resilience building 191
5.5 Systems approach to baseline indicators 198
6.1 Geographic Information System (GIS) overlay analysis producing a risk map 216
6.2 The relationship between disaster impact, uncertainty and risk reduction 221
6.3 Influences on a risk governance cycle 222
8.1 Underlying influences on successful disaster reduction initiatives 251
8.2 From integrated vulnerability to integrated wellbeing 261

Tables

1.1	Examples of disaster impacts on efforts to meet the Millennium Development Goals (MDGs)	18
1.2	International policy events that reflect a progression of development concerns	34
1.3	Broad typology of institutions identified with disaster and development strategies	36
2.1	Viewing disasters from different sustainability perspectives	48
3.1	Estimated economic damage of disaster, compared to disaster death rate and people affected, for three human development levels	96
3.2	Persons of concern to UNHCR by region 2006	108
3.3	Overall estimates of global population displacement	111
3.4	Influences on the condition of displacement	113
4.1	A health and sustainable development perspective of disaster reduction	128
4.2	Health and disasters in Africa	132
4.3	Causes of death for under five years of age in Africa, 2003	132
4.4	Projections for extreme weather events for which there is an observed trend and their potential or likely impacts on health	150
4.5	Climate related risks to health based on integrated health security approach	151
4.6	Food security: some underlying issues and options	155
4.7	Influences on person-specific health outcomes	166
5.1	Learning in disaster, development and disaster risk reduction	202
6.1	Some core influences on reactions to risk	223

Boxes

1.1	Millennium Development Goals	17
1.2	Hurricane Katrina	31
1.3	Hyogo Framework for Action: Summary of Commitments 2005–2015	43
2.1	Defining poverty through the core dimension of capability	72
2.2	Zambezia Agricultural Development Project (ZADP), Mozambique	79
3.1	Capital assets that contribute to human security	101
3.2	Representations of the Burma Cyclone and Chinese Earthquake of 2008	119
4.1	Demographic ageing	126
4.2	Infectious disease in Africa	132
4.3	Sphere standards for health systems and infrastructure	171
5.1	Disaster reduction and schools	182
5.2	Risk and Resilience Committees (RRC)	188
5.3	Institutional guidance on EIA in the UK	193
6.1	Cyclone Preparedness Programme (CPP) - Bangladesh	213
6.2	Late lessons from early warnings	220
7.1	Why do people move back home rather than mitigate risk by going somewhere safer?	232
7.2	The UK Civil Contingencies Act	235
7.3	Sphere standards presented as common to all sectors	242

Acknowledgements

This book contains perspectives on the association between disasters and development. The book is based on a combination of my own thoughts and observations and existing accounts from a wide range of sources. My work with this topic has involved interaction with a committed community of academic and practitioner colleagues. It would be inappropriate to attempt to list everyone, for risk of including some whilst omitting others. However, I would in particular like to acknowledge the staff, postgraduate students (MSc and PhD) and affiliates of the Disaster Management and Sustainable Development programme, and the accompanying Disaster and Development Centre (DDC), at Northumbria University. It was a pleasure to have a chance to lead a programme of this nature from the late 1990s until 2003, as it has been to direct the accompanying DDC since its establishment in 2004. It is encouraging that an increasingly sensitised academic and practitioner community have focused their attention on the disaster and development nexus. The subject is perpetually poised for advancement far beyond what can be captured in this introductory text.

I acknowledge that the linking of disaster reduction and sustainable development activities has unofficially been going on for as long as people have tried to improve the world. The need for solutions to this equation is enormous and in many ways we have only just begun to address them. It should be no surprise and is to be welcomed that some aspects of the relationship between disaster and development have become recognised and institutionalised through international accords, platforms, national, local or independent initiatives. The implications and applications of this way of thinking for the day-to-day plans of both development practitioners and disaster managers are also becoming more evident. There remains a rich field of research. There are opportunities to make a difference in disaster and development from many angles, from beyond and from within academia. This text hopefully represents a contribution to the basic knowledge and understanding that is needed to progress this responsibility, in particular for those more recently entering the subject area.

I would like to thank Andrew Mould, Michael P. Jones and Jennifer Page at Routledge, and Tony Binns the Series Editor, for their recognition of the relevance of the book, and for their patience and support during its production. Many thanks also to three anonymous referees whose comments on the draft manuscript were particularly well informed and appropriate. Thanks also to Lisa Williams for her patience and care in copy editing the text, and to Emma Hart, Production Editor.

The author and publishers would like to thank the following for kindly granting permission to reproduce material: the Department for International Development (DFID) and Crown Copyright, UK for Table 1.1; the Centre for the Study of the Epidemiology of Disasters (CRED) for an earlier version of Figure 1.2; Save the Children for Figure 5.1; Springer Science and Business Media for Figure 6.1; and Janaka Jayawickrama for Plates 1.5, 3.1 and 3.3. Every effort has been made to contact copyright holders for their permission to reprint any duplicate material in this book. The publishers would be grateful to hear from any copyright holder who is not here acknowledged and will undertake to rectify any errors or omissions in future editions.

Abbreviations

AIDS — acquired immunodeficiency syndrome
ALNAP — Active Learning Network for Accountability and Performance in Humanitarian Action
AMREF — African Medical and Research Foundation
AU — African Union
BCM — business continuity management
BCR — benefit–cost ratio
BSE — bovine spongiform encephalopathy
CAMPFIRE — Communal Areas Management Programme for Indigenous Resources Project
CBA — cost–benefit analysis
CBDM — community based disaster management
CCA — Civil Contingencies Act (UK)
CCP — Cyclone Preparedness Programme (Bangladesh)
CFC — chlorofluorocarbon
CRED — Centre for Research on the Epidemiology of Disaster
CSO — Civil Societal Organisation
DAC — Development Assistance Committee (DAC–OECD)
DDC — Disaster and Development Centre
DEC — Disasters Emergency Committee
DEM — digital elevation model
DFID — Department for International Development
DOH — Department of Health
DRC — Democratic Republic of Congo

ECHO	European Community Humanitarian Aid Department
EEA	European Environment Agency
EIA	Environmental Impact Assessment
ENSO	El Niño Southern Oscillation
FAO	Food and Agriculture Organisation
FEMA	Federal Emergency Management Agency
FEWS	Famine Early Warning System
FEWSNET	Famine and Early Warning System Network
FMD	foot and mouth disease
FMO	Forced Migration Online
FRELIMO	Frente de Libertação de Moçãmbique (Front for the liberation of Mozambique)
GATT	General Agreement on Tariffs and Trade
GDI	Gender-related Development Index
GDP	gross domestic product
GEM	Gender Empowerment Measure
GIS	Geographic Information System
GNP	gross national product
GTZ	German International Development Agency
HDI	Human Development Index (UNOP)
HFA	Hyogo Framework for Action
HIV	human immunodeficiency virus
HPN	Humanitarian Practice Network
IASC	Inter-Agency Standing Committee
ICRC	International Committee of the Red Cross
ICT	information communication technology
IDP	internally displaced people
IDRM	infectious disease risk management
IDS	Institute of Development Studies
IFRC	International Federation of the Red Cross and Red Crescent Societies
IGO	inter-governmental organisation
IMF	International Monetary Fund
IOM	International Organisation of Migration
IPCC	Intergovernmental Panel on Climate Change
IRR	internal rate of return
JICA	Japan International Development Agency
KAP	knowledge, attitude and practice
LDCs	Least Developed Countries
LPA	local planning authority
LRF	Local Resilience Forum

MDGs	Millennium Development Goals
MIPAA	Madrid International Plan of Action on Ageing
MSF	Médicins Sans Frontières
NGHA	non-governmental humanitarian agency
NGO	non-governmental organisation
NHS	National Health Service (UK)
NPV	net present value
NSET	Nepal Society for Earthquake Technology
ODA	official development assistance
ODI	Overseas Development Institute
OCHA	United Nations Office for the Coordination of Humanitarian Affairs
OECD	Organisation for Economic Cooperation and Development
PAHO	Pan-American Health Organisation
PHC	primary health care
PRSP	Poverty Reduction Strategy Paper
PVA	Participation Vulnerability Assessment
RCA	resilience capacity assessment
RRF	Regional Resilience Forum
SAD	seasonally affected disorder
SAP	Structural Adjustment Programme
SARS	severe acute respiratory syndrome
SHA	Scottish Health Authority
SIA	social impact assessment
SIDA	Swedish International Development Agency
SMART	Specific, Measurable, Achievable, Relevant, Time-bound
STAC	Scientific and Technical Advice Cell
TB	tuberculosis
UN	United Nations
UNCED	United Nations Conference on Environment and Development
UNDP	United Nations Development Programme
UNEP	United Nations Environment Programme
UNESCAP	United Nations Economic and Social Commission for Asia and the Pacific
UNICEF	United Nations Children's Fund
UNISDR	United Nations International Strategy for Disaster Reduction

UNU–EHS	United Nations University Institute of Environment and Human Security
USAID	United States Agency for International Development
VCA	Vulnerability and Capacity Assessment
vCJD	variant Creuzfeldt-Jakob disease
WCDR	World Conference on Disaster Reduction
WCED	World Conference on Environment and Development
WCSD	World Conference on Sustainable Development
WDM	World Development Movement
WFP	World Food Programme
WHO	World Health Organisation
WTO	World Trade Organisation

1 Introduction: why disaster and development?

Summary

- **There are many ways of interpreting events that may be defined as disasters.**
- **No disaster is 'natural' in terms of the association between disasters and development.**
- **Disasters, disaster reporting, and disaster and development institutions have increased in recent decades.**
- **Disaster prevention and response alter the impact of disasters on development, and appropriate development can reduce disasters.**

Introduction

The background to this book is the consolidation of awareness that what we consider a disaster can be interpreted in terms of development, and that the right type of development reduces disasters. The book explores the reasons why we should focus on disaster reduction and sustainable development as part of the same agenda. It introduces many of the key ideas, terminologies, implications and applications that are part of the interconnected world of disaster and development.

Whilst this book was being written, numerous recognised disasters were reported from around the world. Those that reached the press included the rapid onset events of Cyclone Sidr in Bangladesh on 15 November 2007,

Cyclone Nargis in Burma on 2 May 2008 and the Wenchuan Earthquake of Sichuan Province, China on 12 May 2008. These largely unpredicted examples alone accounted for an estimated 90,000 deaths and about 25 million people being displaced from their houses. Given the level of destruction that occurred, many of these people lost their homes on a more permanent basis. Instability, displacement and economic destitution also continued for many more millions in other parts of the world, including the Democratic Republic of Congo, Zimbabwe, Sudan, Somalia, Iraq and Afghanistan. Across the Horn of Africa, the World Health Organisation (WHO) was estimating by mid-2008 that about 15 million people were facing a combination of drought, armed conflict, high food and fuel prices, and a succession of poor harvests (WHO 2008b). In September 2008, the annual cycle of cyclones were hitting the Caribbean and Southern United States, again reminding people that few areas of the world, whether associated with high, medium or low income, escape environmental hazards. There are many other events that have been dragging on over a long timeframe, such as persistently high infectious disease incidence, development induced displacements (i.e. related to dam, construction, agrarian changes or deforestation) or smaller more localised disaster events that only make it into the local press, or remain unreported. In each instance, a variety of perspectives are used to explain the circumstances that occurred before, during and in the wake of the disaster event.

The causes, impacts and longer-term consequences of disasters are often brought to the attention of an international audience that is experiencing increased connectivity through telecommunications and travel. This occurs alongside personal experiences of crisis, concern about the pace of change we are witnessing, uncertainty about the future and a sense of disaster risk being out of control through climate change. It has led to an upsurge of interest in explanations for disasters, and the means to their reduction. With the emergence of such a wide interest group, one common observation has been that major disruptive events are best approached using varied expertise and knowledge. There have been significant contributions to disaster studies over the decades from disciplines such as geography, environmental studies, economics, sociology, public health and planning. The disaster and development perspective outlined in this book confirms these and links that can be wider still. The field of disaster reduction concerns prevention strategies, political will, community actions, rights, critical infrastructures, survival strategies, relief and recovery, behaviour, perception and health, to name just a few. In this context, we have learnt that supposed ‘expert knowledge' can at times mean a conundrum of missing information and failed expectations. After all, expert knowledge has not been 'expert' enough to prevent the disasters of our times. Increasingly, the role of local knowledge, grounded in local realities, provides a crucial component of the subject area. This is often beyond the reach of the formalised academy and of textbooks. Further interpretational nuances of disaster and development are to be found within ourselves, influenced by our hopes, aspirations and roles in securing wellbeing now and for future generations.

The more formal and extensive disciplinary terrain supporting this subject includes the topics represented in Figure 1.1. A selection of key topics has been included alongside the academic classification with which they might

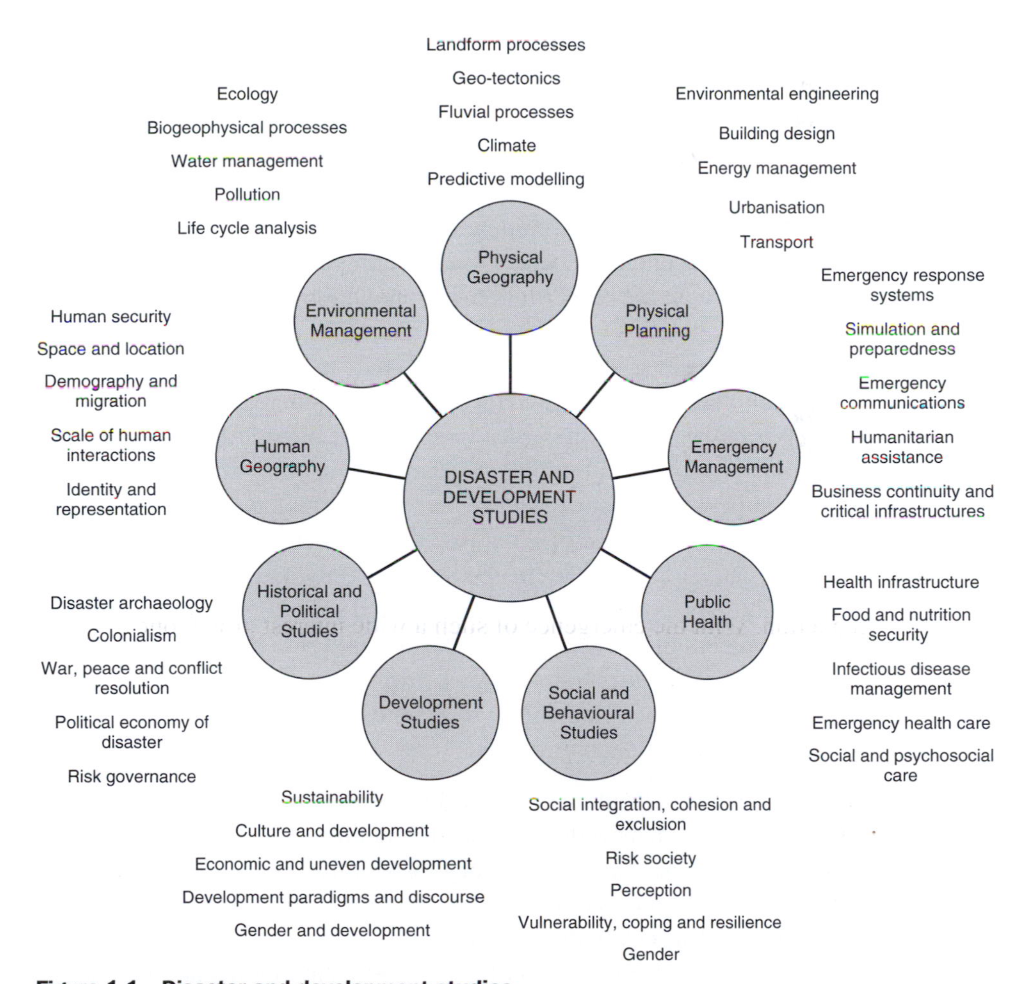

Figure 1.1 Disaster and development studies

be suitably identified. It is recognised that topics in nearly all instances can be situated in more than one of the broad subject classifications.

Overview of the book

The book addresses how we can approach and apply a disaster and development perspective in an integrated way for engaging some of the major crises of our times. Disasters in the context of development are considered to be any severe disruption to human survival and security that overwhelms people's capacity to cope. This definition is broadly in line with some of the more recent policy documents on disaster reduction of international and national development institutions, some of which are referenced in this chapter. Guidance is provided at the end of each chapter on further reading. A key purpose is to demonstrate throughout the importance of addressing disaster and development perspectives together, as two sides of the same issue.

Chapter 2 addresses the disaster and development link from the perspective of how different types of disaster can be viewed from different perspectives of development, including historical and contemporary, economic, social and environmental viewpoints. The chapter starts with the early Malthusian predictions that by this century the world's population would have fouled its own nest beyond the limits of survival. However, development has to date in many ways staved off disaster through technology and through adaptation. Views on basic development theories are presented in terms of their contribution to disaster adaptation or otherwise; through avoidance of disaster, mitigating disaster impact, and post-disaster recovery. This includes an account of disasters and (1) the logic of economic development, (2) the dependency of poorer nations and (3) the link between poverty and environmental degradation. The contexts of urbanisation and globalisation bring new hazards, vulnerabilities and risk governance issues to bear. Different development paradigms coexist rather than replace each other, such that underlying causes of vulnerability to disaster must be interpreted in terms of complex development processes.

Chapter 3 conversely addresses how disasters influence development. Disasters cause extensive loss of life and livelihood, slowing development or setting it on a backwards trajectory for months, years or decades. The typology of disasters is addressed based on their nature and impact on development. Ultimately the definition of disaster is to do with human security. Loss of human security (food, health, environmental, livelihood, social, personal) prevents sustainable development. The process of losing security occurs through loss of capital assets. The chapter introduces these assets (human, social, natural, physical and financial) as being influenced by environmental catastrophes, conflicts and complex political emergencies. The chapter also includes a well-honed sentiment that disasters also result in opportunities. The old adages that adversity brings out the best in people and that necessity is the mother of invention can be exemplified in local responses to disaster. Whilst insecurity is overall bad for development, the reality is that, under pressure, people cope and become resilient, reducing disaster impacts significantly. However, policy environments, such as, for example, those relating to resettlement and refugees, are often not conducive to letting people rebuild their lives, therefore constraining development. Emergency relief and development work is limited if it does not recognise what is meant by local coping and resilience. This raises a deeper problem concerning the categorisation of disaster survivors as helpless victims rather than capable people. Importantly, in responding to disaster issues, representation of a 'disaster' by the media has a significant influence. This can have a direct bearing on when and how aid ends up being delivered, with consequences for development in the recipient areas.

Chapter 4 provides a focus on health disasters, and on health disaster prevention. A chapter dedicated to this topic is not difficult to justify. Whilst disease remains the largest cause of mortality and morbidity in the world, and is increasingly included as a disaster category, different aspects of health (physical and mental) provide both actual and analogous representations of disaster and development. Whilst incidence of disease in an endemic region becomes regarded as the norm, deviation from regular occurrences is considered an epidemic. This leads to subjective interpretations as to when an endemic or epidemic situation might or might not be considered a disaster, a conundrum with resonance across the disaster reduction field. 'Common' diseases of poverty, such as diarrhoea, become 'acceptable' and persist, whilst new disease risks with low death tolls, such as CJD, SARS and bird flu, are considered a major disaster threat and worthy of significant financial investment. The human immunodeficiency virus and accompanying acquired immunodeficiency syndrome (HIV/AIDS) provide a poignant example of where disease disasters hold back development, such as through the death of high percentages of the economically active age group in sub-Saharan Africa. However, strategies aimed at better living with HIV/AIDS, with the use of antiretroviral drugs, are beginning to spread hope of longer-term recovery from the effects of this disease on society.

Well over half of infectious disease incidence is associated with food and nutrition security, the precursor to famine disasters and ultimate indicator of insecurity and underdevelopment. This is usually aggravated by drought rather than being a direct consequence of it. With global calorie supply outstripping demand in recent years, famine should be considered an avoidable slow onset disaster, the progression of which is symptomatic of the failings of uneven development. Psychosocial influences accompany all disasters, but are often neglected or misunderstood. Biomedical and more culturally based approaches to addressing trauma contrast with each other. Whilst primary health care (PHC) is a mainstream of development, its basic principles get modified for health care in emergencies. There are direct links between preventative health care, disaster risk reduction and environmental care, and the ideology of primary health care translates closely to current disaster risk reduction thinking.

The three chapters that follow focus on applied aspects of the topic, familiar to the experience of many disaster management and development practices. They address the learning and planning process in disaster management (Chapter 5), early warning and risk management approaches and techniques (Chapter 6), and disaster mitigation, response and recovery (Chapter 7).

Chapter 5 represents a 'learning by doing' approach, exemplified by the project cycle and, in the wider sense, by an interpretation of the disaster management cycle. Review and evaluation of development are paralleled to some extent by the process of review and lessons learnt post-disaster. Both processes are arguably dependent on information based on local knowledge, and experiences of development and disaster, if they are to progress effective hazard management and vulnerability reduction. The chapter presents data collection approaches broadly described as needs, vulnerability and capacity assessments.

Planning interventions also requires identification of the roles of different stakeholders and the systematic representation of plans to funders and implementers. Impact assessments used for planning development work include a risk assessment component, which is fundamental to a disaster reduction approach. The application of review, monitoring and evaluation for both work with disaster reduction and development is also covered. This is presented in a way that contributes to an undercurrent throughout the chapter, that disaster reduction and development policy should be guided by practice and be responsive to real demand. It is not suggested that this is simple to achieve. However, application of more of a precautionary principle in disaster and development work is suggested, particularly in interpreting the output of risk assessments. Ultimately, learning and planning for disaster reduction involve everybody, not just project teams, such that education curricula for schools need to engage the next generation in what to do and how to interpret the complex world of disaster and development.

The approaches to and techniques for disaster reduction and development continue into Chapter 6 with an account of the background and current status of early warning systems and risk management. These are dependent on the selection of indicators, agreement on thresholds of risk, and minimum standards in making a response. Some of the official guidelines are flagged in this section, such as the Sphere guidelines, but with remaining questions about their applicability in varying disaster and development situations. Environmental, social and economic surveillance approaches and techniques draw on famine early warning and technological drives to predict rapid onset events, such as hurricanes, earthquake and tsunami. Early warning information is used for predictive models, but these raise inherent problems about predictability and uncertainty in complex systems.

Ultimately, we face major challenges in dealing with the interaction of environmental uncertainty and unpredictably in human systems. This is a complex field that development has not managed to keep up with. More development presents greater demands for reducing uncertainty, whilst representation of risk and perceptions remains notoriously subjective. From choosing to live on a flood plain, to unprotected sex, smoking and polluting the atmosphere, research demonstrates a wide variety of reasons why some people take greater risks than others. Different ways of assessing risk are provided. Early warning and risk assessment lead to risk management that can be applied in a developmental way when potential disasters are less evident. Appropriate contexts within which to reduce disaster risks depend on there being an appropriate system of governance and the political and personal will to act.

Chapter 7 focuses on the responses that are made in disaster management and how they are variously divided between the practices of preparedness, avoidance, impact reduction, responding with emergency relief, and recovery. Reponses are generally interpreted as humanitarianism at all levels, local, national and international, but are motivated by different traditions and beliefs. A further response to disaster and the threat of disaster is to develop more resistant infrastructure, such as through earthquake resistant building initiatives (e.g. in Japan and Nepal) and cyclone shelters (e.g. in Bangladesh). However, there are many examples of where development has produced a more susceptible infrastructure, such as through expansion of dense human settlements onto unstable hill slopes, or clearances of swampland and mangrove. Preservation of the latter would have better protected some of the victims of the Indian Ocean Tsunami.

Much of disaster resilience is related to institutional strengthening. Poor governance predisposes to increased impacts of disaster, or can directly cause a disaster, such as when building regulations are not in place, investments are not made in preparedness for emergencies or where aid and relief are corrupted. The antithesis to this is good risk governance in which human security is prioritised and where investment in disaster avoidance saves many times the expenditure that would be required for disaster response. There is increasing recognition of the need to build avoidance and mitigation strategies, not only into disaster prevention, but also into the response and recovery equation.

The final chapter (Chapter 8) provides overall concluding comments on the progress and prospect for sustainable disaster reduction and development. It reiterates the relevance of integrating disaster management and sustainable development thinking as a priority development perspective. It is concluded that investment at all levels, not least the individual, can be made in disaster risk reduction. As the awareness and capacity in the field of disaster and development increase, the expectation would be that a mentality of just coping with catastrophe, or 'survival', would be more about poverty removal, wellbeing and the rights of future generations to live securely.

This opening chapter continues by further introducing some of the terminologies, approaches and rationale to focusing on disaster and development in an integrated way. It is intended that by the end of this book this will have in turn provided the basis for improving our understanding of the potential for disaster reduction, to improve development, human security and wellbeing.

Disaster trends, reporting and definitions

No overview of current trends and statistics for disasters is entirely accurate, as they are reported in different ways from more and less rigorously monitored regions based on definitions of 'disaster' that mean different things to different people. Writers on disasters have frequently

Plate 1.1 The cyclone defence wall at Chakoria, Bangladesh

Source: Author.

Plate 1.2 Landslide remediation in Nepal

Source: Author.

Plate 1.3 Remains of a residence in Mozambique following coastal erosion and storms

Source: Author.

Plate 1.4 A bombed bridge in Mozambique being used to dry clothes

Source: Author.

made the point about there being different ideas about what constitutes disaster, such as the suitably titled text by Quarantelli (1998) on *What Is a Disaster? Perspectives on the Question*. Differences in perspectives on what constitutes disaster are dependent on a sense of loss which varies from one person to the next. Altering the status of disaster can also be a result of politics and humanitarian aid (Middleton and O'Keefe 2001), underreporting of isolated cases, and the uneven interests of the media. Nonetheless, international organisations, such as the International Federation of Red Cross and Red Crescent Societies (IFRC), the United Nations International Strategy for Disaster Reduction (UNISDR), United Nations Development Programme (UNDP), the Department for International Development (DFID), and other regional or national bodies need to calculate changing disaster trends based on approximate common understanding over definitions of disasters. There is broad consensus that disaster includes disruption beyond people's capacity to cope.

Summaries of changing trends in disaster incidence can be obtained from IFRC, which produces an annual *World Disasters Report* (Figure 1.2). This provides data for the total number of people reported killed by disasters by country, alongside those reported as being affected. Similar information can be obtained via UNISDR, a unit established within the United Nations since 2000. Much of the broader data being used is drawn from an overview of disaster trends maintained by the Centre for Research on the Epidemiology of Disaster (CRED) at the Catholic University of Louvain, Belgium, which manages an easy to access database called EM-DAT. In 2007, CRED recorded 414 natural disasters and the fact that 'these killed 16,847 persons, affected more than 211 million others and caused over 74.9 US$ billion in economic damages' (CRED 2008). Figure 1.3 represents EM-DAT in 2007, showing how the reporting of disasters has increased in recent decades. The increase is not just due to better systems of compiling data, though that has an influence on this trend. Rather, we should consider the overall increases in world population that accompany a greater opportunity for more people to be caught in the wrong place, at the wrong time, without adequate forms of protection. The current trend is also, however, partly reflective of an increased number of disruptive events, in particular those of a hydro-meteorological nature, such as floods and storms.

The varied data categories on disasters are also an indication of how they are recognised in different ways. The IFRC is a body that typically has used a classification of 'disasters with a natural or technological trigger' and *not* the 'effects of war, conflict-related famine, disease or epidemics such as HIV/AIDS' (IFRC 2004: 161). However, it is a sign

Figure 1.2 The *World Disasters Report* is an annual release of the International Federation of Red Cross and Crescent Societies

of the changing attitude to disaster classification that the 2008 *World Disables Report* focuses on HIV/AIDS. Meanwhile the remit of the sister organisation, the International Committee of the Red Cross and Red Crescent, specifically includes a mandate for conflict related issues. As will be indicated in Chapter 4, on health aspects of disaster, data for the incidence of infectious and chronic degenerative disease outnumber on a scale of hundreds to one the deaths indicated under the traditional Red Cross disaster categories. However, even taking into account the various limitations of global data, these consistently confirm a steady increase in disasters with a 'natural' or technological trigger. Whilst Figure 1.3 suggests a huge increase in disasters in recent times, some account must be taken here of underreporting of disasters in previous years.

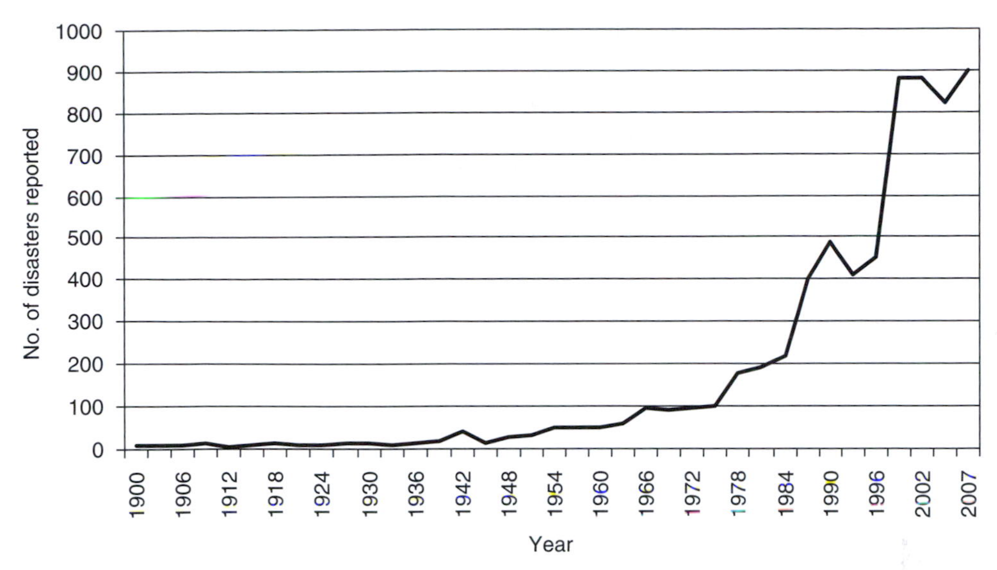

Figure 1.3 Total number of disasters reported 1900–2006

Source: Based on approximate data from EM-DAT CRED (2006).

The same sources of data on disasters also supply information by more detailed disaster type. A typology of disaster types is a significant component of the contribution of earlier authors on disasters and disaster management. They are well covered by those who have emphasised more specifically 'natural hazards', such as Smith (2001). The terminology of natural hazard leads to the notion of a natural disaster, which comes from a tradition in disaster studies that generally focuses on physical environmental events. A natural hazard has essentially been considered to be all those aspects of the physical environment that can cause us harm, disaster being when people are adversely affected. Thus, an earthquake occurring in a remote part of the East African Rift Valley, where there are no people living in buildings or motorway flyovers likely to collapse, would not be a disaster, but conversely the same strength of tectonic event in a built-up area would be. This example also suggests that where there is little or no development, events that may cause disaster in some locations are less relevant as human survival and coping is not put at risk. However, immediate loss of life alone is not the sole criterion for defining disaster in much of this field. Whilst death accompanies all disasters and life in the longer term, the emphasis here is extended to the loss of development potential caused by disasters. This way of defining disaster, whether related to development in the short term or the long term, is a core rationale of the book.

A well-used approach to the understanding of disasters is to look at hazard and vulnerability together, as detailed by Blaikie *et al*. (1994) and further revised for Wisner *et al*. (2004). These works reflect the emphasis within disaster studies of vulnerability to hazards and variable human capacity to mitigate their impact. Whilst disasters are commonly considered events that catastrophically affect people, focusing on human vulnerability presents an important distinction in recognising the breadth of causes and origins of disaster. The quest to interpret disaster in terms of vulnerability is fundamental. Oliver-Smith (1996) points towards three varieties of socially and technologically produced conditions of vulnerability:

1. behavioural and organisational – concerning decision making and systems;
2. social change – concerning societal resilience and values;
3. political economic and environmental – concerning exposure to hazards for historical-structural reasons.

The application of interpretations of vulnerability in a practical sense is an ongoing and vibrant field that can bring hope to the future success of disaster reduction strategies. Changes in vulnerability are influenced by development, including through the accompanying processes of environmental change, conflict, technological mishap, health, politics, culture and so forth. Understanding the relationship between vulnerability and human needs, and how these can be addressed by building resilience to disaster, is not only an important theme within this book, but also part of the recently reiterated global disaster reduction agenda. It is part of the agenda to reduce poverty reduction and environmental degradation, to increase sustainable development, and to advocate suitable systems of governance to accompany these.

Using a disaster and development approach, it is appropriate to consider disaster not as 'natural' but as a function of development. A disaster is recognisable as the consequence of there being insufficient development of a means to avoiding a human crisis, or as an aspect of development itself having been the cause of the crisis. Disaster occurs through exposure to an adverse hazard: being in the 'wrong place' at the 'wrong time' with inadequate forms of protection. The rationale here is that in as much as an earthquake, tsunami, hurricane or flood might be part of nature, the process of human development has not adapted sufficiently to avoid crisis. This view of disaster is part of the response to the vulnerability perspective of disasters referred to above and is a recurrent theme throughout the book. The vulnerability, resilience or wellbeing approaches to disaster reduction extend beyond a sole concern with being exposed to disastrous events, to people's socio-economic status, knowledge, attitudes, physical and mental wellbeing. More importantly, it is demonstrated that vulnerability can be counterbalanced by resilience and coping with disaster as part of development. Ultimately, it is argued that there are real development choices to be made in terms of disaster prevention, including the chance to shape human behaviour and the institutions that govern. There are options to pursue a combined disaster reduction and sustainable development agenda, such that those with this understanding gain the required optimism concerning what could be achieved.

As is explored in more detail in Chapter 2, the history of ideas about disasters reflects some highly contrasting perspectives. Amongst the contemporary ideas, at one end of the spectrum it has been considered from a Malthusian view of the world that human populations are 'naturally regulated' by disaster. These are views common to some studies of population that are still referred to. However, it is debatable whether current statistics prove that the world is disastrously overpopulated beyond its carrying capacity. This is not least because it is known from cases in many parts of the world, including Europe, that human populations regulate themselves through development and not through disasters. This simple logic of development as the regulator of disaster lies at the interface of understanding the means to achieving disaster reduction. With more precise reference to where disaster meets development and vice versa, it can therefore be emphasised that it makes little sense to accept any disaster as 'natural'. Furthermore, care in defining the correct origins and consequent natures of different types of disaster is crucial to taking the right development routes to disaster prevention and control. Over the last thirty years and more we have referred to this process more broadly as 'sustainable development', as the means to saving the planet and its people. In this context a large part of the purpose of sustainable development is synonymous with the aims of disaster reduction.

Development goals and disasters

Current development goals should be born out of recognition of the need to remove the burden of poverty. In achieving this, nation-states and citizens would achieve a more equitable and less risky world. Concerns about human and environmental degradation are reflected in a passage of global events and international accords emergent since the end of the

Second World War. These are returned to later in this chapter in the context of the disaster and development industry. However, there have been other similar occurrences since before the Second World War, such as for example, the British Government's establishment in its colonies of 'famine codes' that reflected a concern about environmental and life sustainability. Increasingly recognisable in the contemporary process is a coordinated international dialogue to try to plan for a world that will allow life to survive indefinitely. This lies at the heart of sustainable development, which has many definitions and is a recurrent theme of the Routledge 'Perspectives on Development' series, of which this book is a part. Perhaps no definition has been used more often to define sustainable development than that of the Brundtland Commission of 1987, which states that 'sustainable development is development that meets the needs of the present without compromising the needs of future generations' (WCED 1987: 1).

Sustainable development has environmental, social and economic dimensions (Hatzius 1996), such that unsustainable development includes the risk of disaster through environmental degradation, social decay or economic collapse, to name a few. Realisation of the consequences of unsustainable development and the current state of persistent mass poverty and human suffering has led to the need for goals and indicators of sustainability (Bell and Morse 1999). This generally happens at the level of local development projects and national development plans. However, more recently it has also been reflected in the establishment of the Millennium Development Goals (MDGs), which were born at the World Conference on Social Development at Copenhagen in 1995 and consolidated at the World Conference on Sustainable Development in Johannesburg in 2002 (Middleton and O'Keefe 2003). There are reasons why major quantitative targets such as the MDGs have so many limitations, including accusations of misrepresentation of what is possible over a short time span, and gimmick, in that pledges are made but actions are insufficient (Saith 2006). However, in terms of concerns identified, the eight goals and the targets to achieving them can also be considered as essentially comprising a large portion of the right agenda for international disaster reduction. They are listed in Box 1.1.

In a development view of disasters it is important to note that disasters may be either 'slow onset' or 'rapid onset'. There are examples of both that are associated with development. For example, if we consider goals 1 and 7, on poverty and environment respectively, we might think of where underdevelopment causes people to construct makeshift housing on steep slopes or hazardous parts of the flood plain, as can be witnessed in many

Box 1.1

Millennium Development Goals

The eight goals are as follows:

1. eradicate extreme poverty and hunger;
2. achieve universal primary education;
3. promote gender equality and empower women;
4. reduce child mortality;
5. improve maternal health;
6. combat HIV/AIDS, malaria and other diseases;
7. ensure environmental sustainability;
8. develop a global partnership for development.

The targets include:

1. between 1990 and 2015, halve the proportion of people whose income is less than US$1 a day;
2. reduce by two-thirds, between 1990 and 2015, the maternal mortality rate;
3. by 2015, have begun to reduce the incidence of malaria and other major diseases;
4. halve, by 2015, the proportion of people without sustainable access to safe drinking water and basic sanitation.

Source: United Nations (2008).

parts of the developing world. Through this, people become more exposed to the effects of flash flooding, a rapid onset disaster. However, where land degradation over many years reduces the land productivity for food on which people subsist, a slow onset disaster unfolds as food insecurity moves closer to conditions of famine. Poverty, health, education, human rights, environment and good governance lie at the heart of the MDGs and these are also what lies at the core of reducing the risk of disaster.

A series of further examples of disaster impacts on efforts to meet the Millennium Development Goals is provided by the Department for International Development policy document on *Reducing the Risk of Disasters* (DFID 2006) and is included as Table 1.1. In each instance, the direct impacts of disasters have been presented alongside the presumed indirect impacts. This summary of key issues to consider provides a good orientation to thinking about the circumstances within which disaster reduction might contribute to achieving the MDGs.

Table 1.1 ***Examples of disaster impacts on efforts to meet the Millennium Development Goals (MDGs)***

MDG	*Direct impacts*	*Indirect impacts*
1 Eradicate extreme poverty and hunger	• Damage to housing, service infrastructure, savings, productive assets and human losses reduces livelihood sustainability.	• Negative macroeconomic impacts, including severe short-term fiscal impacts on growth, development and poverty reduction. • Forced sale of productive assets by vulnerable households pushes many into long-term poverty and increases inequality.
2 Achieve universal primary education	• Damage to education infrastructure. • Population displacement interrupts schooling.	• Increased need for child labour for household work, especially for girls. • Reduced household assets make schooling less affordable – girls probably affected most.
3 Promote gender equality and empower women	• As men migrate to seek alternative work, women/girls bear an increased burden of care. • Women often bear the brunt of distress 'coping' strategies, e.g. by reducing food intake.	• Emergency programmes may reinforce power structures which marginalise women. • Domestic and sexual violence may rise in the wake of disaster.
4 Reduce child mortality	• Children are often most at risk, e.g. of drowning in floods. • Damage to health and water and sanitation infrastructure. • Injury and illness from disaster weakens children's immune systems.	• Increased numbers of orphaned, abandoned and homeless children. • Household asset depletion makes clean water, food and medicine less affordable.
5 Improve maternal health	• Pregnant women are often at high risk of death/injury in disasters. • Damage to health infrastructure. • Injury and illness from disaster can weaken women's health.	• Increased responsibilities and workloads create stress for surviving mothers. • Household asset depletion makes clean water, food and medicine less affordable.
6 Combat HIV and AIDS, malaria and other diseases	• Poor health and nutrition following disasters weakens immunity. • Damage to health infrastructure. Increased respiratory diseases associated with damp, dust and air pollution linked to disaster.	• Increased risk from communicable and vector borne diseases, e.g. malaria and diarrhoeal diseases following floods. • Impoverishment and displacement following disaster can increase exposure to disease, including HIV and AIDS, and disrupt health care.

Table 1.1 ***Examples of disaster impacts on efforts to meet the Millennium Development Goals (MDGs)—cont'd***

MDG	*Direct impacts*	*Indirect impacts*
7 Ensure environmental sustainability	• Damage to key environmental resources and exacerbation of soil erosion or deforestation. Damage to water management and other urban infrastructure. • Slum dwellers/people in temporary settlements often heavily affected.	• Disaster induced migration to urban areas and damage to urban infrastructure increase the number of slum dwellers without access to basic services and exacerbate poverty.
8 Develop a global partnership for development	• Impacts on programmes for small island developing states from tropical storms, tsunamis, etc. • Impacts on commitment to good governance, development and poverty reduction – nationally and internationally.	
All MDGs		Reallocation of resources – including official development assistance (ODA) – from development to relief and recovery.

Source: DFID (2006).

Introducing themes that support the disaster and development nexus

This section looks in more detail at some of the themes and frameworks that can be used in analysing disaster and development.

Hazards, vulnerability and capacity

There are several underlying frameworks that can be used in approaching disaster and development. One approach already alluded to is to assess hazards in relation to human vulnerability. This is represented in much of the work on disasters over the decades and is now considered fundamental to understanding disaster management. Thus, Cuny (1983) implied this perspective in his Oxfam book on disaster and development by highlighting the development contexts within which most of the more serious disasters were occurring. The emphasis in these instances is put on how disaster vulnerability is managed by development, and development by disaster prevention. In as much as hazards vary in scale and intensity depending on where, when and with whom they occur, measuring disaster requires ongoing quantification and qualification of its impact. For example, measuring earthquake hazard has looked at, on the one side, the intensity of earth tremors using the Richter scale and, on the other, the intensity of its impact on inhabited places using the Mercalli scale (K. Smith 2001).

Studies of health disasters have similarly addressed the risk of a health disaster in terms of pathogenic hazards versus the risk of a health disaster through human vulnerability to being infected (Collins 1998; Collins *et al*. 2006). The risks, which are ecological and social, have been shown to have a complex mix of origins. Chapter 4 is dedicated to health disasters, which are by far the most extensive of the various categories of global crises. The net result of understanding people's vulnerability to a hazard is that it can be mitigated by concerted interventions. Intervention in disaster events may be driven by many development rationales, which are managed by '*gatekeepers*', or more simply those who make decisions on how to reduce the chance of a disaster. Arguably, all people who have an influence, and some benefit from disaster prevention, are stakeholders in the disaster and development nexus, and also gatekeepers. They include governments, international organisations, businesses, communities, non-governmental organisations, households and individuals. The extent to which any of these can intervene is a measure of disaster reduction *capacity*. The inclusion by Wisner *et al*. (2004) of capacity in their version of a hazard and vulnerability equation is summarised as:

$$\text{Disaster risk} = \text{Hazard} \times \text{Vulnerability/Capacity}$$

Capacity here is synonymous with *capability*, which is core to assessing progress towards poverty reduction. The *World Development Report* of the UNDP uses measures of capability to mean the degree to which poverty is being reduced, overlapping with the aims of the Millennium Development Goals. There are a multiplicity of assets that could be measured to describe capability or otherwise. To monitor development projects and programmes, key indicators represent changes in progress towards targets. For example, on the wider scale of the nation-state, a capability index has been used and has been included in the UNDP Annual Report in various forms since 1997, as follows:

- capability to be well nourished and healthy – represented by the proportion of malnourished under-5 children;

- capability for healthy reproduction – proxied by the proportion of births unattended by trained health personnel;
- capability to be educated and knowledgeable – represented by female illiteracy.

(UNDP 1997: 16)

This contrasts with other development indexes used by UNDP, such as the *Human Development Index*, which includes income measured through gross domestic product (GDP) per capita alongside life expectancy and educational attainment. It also contrasts to some extent with the *Human Poverty Index*, which looks at the percentage of people expected to die before age 40, illiterate adults, people without access to health services and safe water, and underweight children of less than five years of age (UNDP 2002 is more detailed). The difference in versions of indexes that refer to capabilities is that they draw attention to both people's capacity to intervene through knowledge and education and also their access to the means of intervention through basic human rights. This development of the concept of capability lies at the heart of much of the development and disaster reduction discourse and is repeatedly revisited in subsequent chapters.

If capability is not just about education and the presence of the benefits of development, but also rights, then we must consider many disasters as also being fundamentally about systems of governance and the state of human societies. During conflict, people, communities, cultures, environments and economic systems are put at risk. Often the threat from an environmental hazard is accentuated by conflict, such as in the case of the mass famines in the Horn of Africa during the last three decades (Devereux 1993; Devereux and Maxwell 2001; De Waal 1997). Conflict prevents the adaptation of basic livelihoods survival, destroys the possibility of alternative trade that can benefit local people and cuts off supply lines for assistance. The politics of humanitarian assistance leaves whole populations dependent on others and susceptible to further disaster and underdevelopment (Middleton and O'Keefe 2001). The link between an environmental hazard and conflict was also realised in the case of the tsunami in the Indian Ocean at the end of 2004. Amongst other locations, the tsunami hit the eastern coast of Sri Lanka and the northern province of Aceh, Indonesia, both locations with rebel movements fighting for autonomy. Whilst a catastrophic environmental disaster may have the effect of drawing people in conflict together, initially witnessed in these regions, control over the delivery of assistance can encourage persistence of community and political divisions. Where there is no resolution of the

Plate 1.5 Tsunami damage of a community centre in Sri Lanka
Source: Janaka Jayawickrama, Northumbria University.

underlying issues that divide people, there is less hope of building capability to deal with the threat and onset of disasters. The link between disaster severity and human rights cannot be overestimated. This seems to have been realised in Sri Lanka, where the world's first Ministry of Disaster Management and Human Rights was established in 2005.

Multilevelled and systems analysis approaches

Hazards, vulnerability and capability can be assessed at the global, regional, local community, household or individual level, something we refer to as a *multilevelled approach*. Multilevelled approaches are used in many types of geographical and economic studies. It is the idea that processes of change in the environment, society or the economy happen at different levels which interact with each other. Examples in environmental management are provided by Wilson and Bryant (1997) and, by way of example in a regional study, by Abrahamsson and Nilsson (1995). Figure 1.4 demonstrates that interaction between these different levels is part of development and change over time. In terms of disasters and development, we might consider this very relevant in terms of the adage that there are local solutions to global problems, so think global

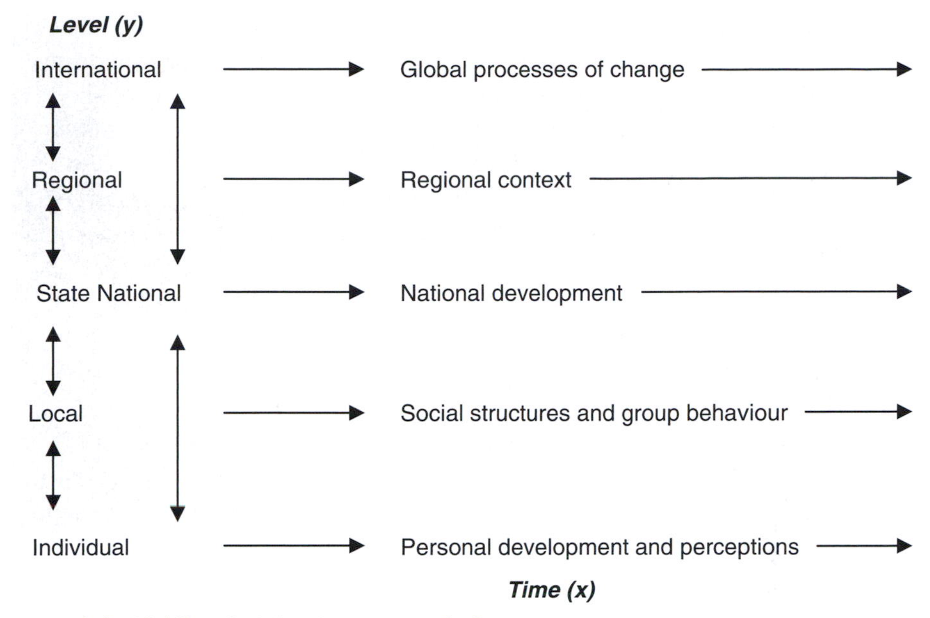

Figure 1.4 Multilevelled development analysis

Source: adapted from Abrahamsson and Nilsson (1995).

and act local, but also in terms of how *globalisation* alters the risk of disasters at the local level.

Interaction between the causes of disasters at the global level and those at the local level is part of a *systems analysis* approach. In this instance, environmental, social and economic changes may collide in contexts of disaster represented as the flows of different disaster risks during development. For example, in rural areas the climatic cycle of rainfall, potential drought and flood is synchronised with economic investments in planting and change in day-to-day livelihood activities. People in crop producing areas have adapted their cultivating systems to these regular changes. However, where the regular cycles of rainfall and flood become more randomly large or deficient, there is an impact on normality throughout this adapted *livelihood system*.

Systems approaches in this context require an understanding of the environment and resultant attitudes to it as being situated in a theoretical framework of more and less deterministic ideas. These are identified as ranging from crude *environmental determinism* to those ideas grounded in *phenomenology*. Environmental determinism is the view 'that the environment controls the course of human action' (Lewthwaite 1966, cited in *The Dictionary of Human Geography*). For example,

Charles Darwin's theory of evolution is deterministic. It proposes that organisms evolve in relation to the environment, that competition from other living organisms causes evolution. Meanwhile, phenomenology recognises 'observation' and 'objectification' as not a simple form of deducing facts from conventional forms of science, but a philosophy which seeks to understand the world as it shows itself before undergoing scientific inquiry. It is an alternative to *positivism*, which is a philosophy originally proposed by Auguste Comte in the 1820s and 1830s to distinguish science from meta-physics and religion (Pepper 1984, 1996). Features of positivism are to be concerned with (1) the direct and immediate experience of the world, (2) advocating a more unitary scientific method that should be accepted and drawn on by a wide scientific community and (3) the construction of theories capable of empirical (scientifically proven) verification. Meanwhile, *possibilism* is the view that physical environments provide the opportunity for a range of possible human responses and that people have considerable discretion to choose between them. For example, Barr argues that 'environmental action is open to a range of influences, focusing especially on environmental values, situational characteristics and psychological variables' (Barr 2003: 227). Consequently, Larson and Lach (2008) have shown how in Portland, Oregon, a variety of groups needed to be considered in engaging varied attitudes to participation in water resources management.

Note that these are much simplified descriptions of the implications of determinism and possibilism, but they indicate how there can be more and less deterministic ideas about how we engage with the environment. It is the rationale encapsulated in more deterministic or possibilistic interpretations that fundamentally influences the way we approach issues in disaster and development. Furthermore, theories and frameworks influence the ways in which all phenomena are perceived, observations being interpretations of observed facts in the light of theories (after Popper 2002). An implication is that an improved theoretical understanding of environment, development and disaster should ultimately guide an improved engagement of both individual and institutionalised approaches.

Processes of change from a systems perspective are part of a wider whole that helps us to understand the influences on complex phenomena. Early systems analysis developed out of the technical world of electronics and was centred on closed systems, in which everything connects to everything else in a circuit of activity. Further application of systems analysis to the environment is associated with our improved understanding of *ecosystems*. There are many examples of these in basic ecology texts, such as, for example, the carbon, water or nutrient cycles, which are crucial to maintaining balance in the earth's life systems. The other type of system that can be applied to environmental thinking is the open system. The rationale for describing processes in the environment more in terms of an open system is that closed environmental systems leave many unanswered questions concerning the nature of environmental complexity. Qualitative information about the earth's landscapes is also valid in its interpretation and can be allowed for in an open system. More open systems analysis tends to conceptually emphasise human interactions with an environmental system, with less predictable outcomes. Even without human inputs, aspects of the natural physical environment can also be considered to have open-ended incalculable elements, particularly since we have become more aware of some of the more chaotic aspects of natural systems.

Structure and human agency in disaster and development

It was suggested in the previous section that advocates of 'good science' are often prone to determinism in that effects or outcomes are linked to direct causes. Typically, a larger problem is broken down into small parts so that individual links can be explained. This is known as a *reductionist* approach. For example, a direct link between the productivity or survival of key crops in relation to the presence of nutrients and water might be scientifically established. Controlled laboratory experiments can test a crop variety under varying degrees of nutrient deprivation. This represents a legitimate experiment to verify specific hypotheses about the tolerance of the crop. However, in the complex world of human interactions with varying environments and with uncertain changes and predictions, it is often inadequate to directly associate cause and effect. Despite a relationship between different phenomena, one does not necessarily *cause* the other. For example, urban areas of higher than average population density may have higher than average incidence of accidents. However, it is not necessarily the high density of the population *per se* that causes the higher incidence of accidents. A further explanation might be that those high density areas have a high incidence of poverty or are life-threateningly stressful places to live in, independent of population density. There are likely to be complex inter-associations between different influences. With this example we would need not only to understand these inter-associations, but also to gain a better understanding about the way people live in terms of *human behaviour* and the social and economic environments within which they must survive (*structural context*).

The capacity of human beings to decide is often referred to as *human agency*. Human agency is also a function of what people want to do and what they are forced into doing by their circumstances. We can call the organisational context of human institutions and their power to influence people's lives *structure*. There is therefore often a distinctly political aspect to the term. For example, Marxist ideas are essentially structural in emphasising that it is the power of colonially oriented governments and the capitalist system that has created underdevelopment in much of the world. More and less structural approaches can also be recognised within emergency management systems. For example, Waugh and Streib (2006) discuss the structural approach that came to United States' emergency management with the creation of the Department of Homeland Security. This imposed a more top-down and structured command and control approach, conflicting with that of the pre-existing Federal Emergency Management Bureau (FEMA), which had adopted a more collaborative organisational culture.

Whilst a more sociological view of environment, development and disaster does not deny independently driven environmental forces, it addresses how this is perceived by different social entities and, importantly, how their perception of its importance affects their action towards it. This is called *social constructionism* (Hannigan 1995). The entire discussion about how we look at cause and effect in disasters in terms of both structure and agency can be reviewed in terms of *post-structuralism*. This is where disaster and development may be considered as a more relative concept acknowledging the real or perceived nature of disaster and underdevelopment dependent on varied disciplines, cultures and means to understanding. It allows for a variety of meanings in the descriptions of disasters that are not reduced to singular political or cultural perspectives. For example, whilst a structural political economy of disaster might be about how politics and uneven development have made people more prone, a post-structural view might extend to a wider set of possibilities regarding the relative interpretations of disaster and its causes. The field of disaster sociology explores the realm of our conceived notions of disaster. A more detailed debate on its emergence and significance can be accessed via the papers of Dombrowsky (1998) and Hewitt (1998), amongst others, in Quarantelli (1998).

The disaster management cycle

Turning more specifically to disaster management frameworks, a progression of disaster related activities is often considered in terms of

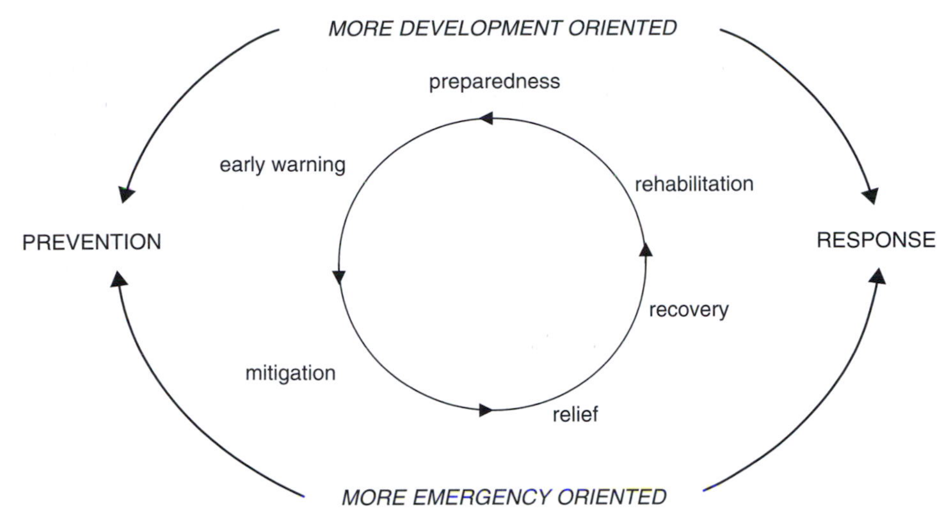

Figure 1.5 Disaster management cycle and development

preparedness, early warning, mitigation, relief, recovery and rehabilitation phases (Figure 1.5). These steps in addressing disasters are usually referred to as the *disaster management cycle*, the stages of which are returned to in some detail in future chapters. Analysis of different representations of the disaster management cycle can also be accessed in work by Frerks *et al.* (1995) and by Von Kotze and Holloway (1996). The first three steps in the cycle used in Figure 1.5 are generally thought of as the preventative actions and the second three as the response actions. All of this process is 'developmental', as lessons learnt from one disaster are supposedly applied to improved prevention and relief for further disasters. Different texts have used the terminology in varying ways. For the purposes of this book, *preparedness* means being ready for disaster through having an adequate level of development in disaster reduction. *Early warning* is being able to predict a disaster event and to ensure that people who are at risk are aware of what is going to happen. *Mitigation* is reduction of the impact of potential disaster events up until and whilst they are occurring. *Relief* is the response actions that reduce the impact of the disaster event after it has happened. *Recovery* is the process of restoring lives, livelihoods and infrastructure to a locally acceptable standard. *Rehabilitation* is dealing with the longer-term effects of a disaster and a fuller restoration of development.

All stages of the disaster management cycle require sustainable development solutions if disaster risk is to be reduced. However, in the sustainability context, the rehabilitation, preparedness and early warning stages might be considered more as development actions. Mitigation, relief and recovery are more frequently part of emergency management. However, as there are multiple disaster risks, and any one of a number of changes may induce a critical event, it is better to consider cycles within cycles such that all of these actions are potentially required simultaneously and in close proximity. As such, development may be both part of prevention and also part of response, and the emergency phase may include preventative actions to stop a current disaster giving rise to more. The key point about this staged approach to disaster management is that the right type of development in the pre-disaster stages can prevent a disaster and that the response to a disaster can be part of the prevention of further disasters. Appropriate development can provide the means to avoid disasters, mitigate their impact or aid in sustainable recovery once one has occurred. As such, there is a need for both development and emergency oriented prevention and response.

Development induced disaster

Dependent on changing world contexts and human choices throughout the ages, development has either reduced or increased human vulnerability to disaster. The quest to understand the circumstances in which different development paradigms can induce or reduce disaster risks is therefore a key part of disaster management. Some of the development paradigms are touched upon in this book, particularly in Chapter 2. A very lucid overview is provided by Willis (2005) as a part of this series. Understanding development as a part of dealing with disasters has historically been a badly neglected field. This is in part due to the twin fields of sustainable development and disaster management having emerged as separate entities. In the context of the developing world, a bridge between development and relief policy and practice amongst the disaster and development agencies has, however, in many instances already now been articulated.

It has been stated in this opening chapter that vulnerability to disaster may be through underdevelopment, defined as a situation in which poverty creates both social and economic susceptibility. Vulnerability can, however, also be induced by development, which can increase marginalisation, and loss of the means to cope with disasters. In the case of technologically and economically developed contexts, society has sometimes increased its own vulnerability through technological dependence or by being locked into a system of disaster provocation.

This would be the view of many concerning the contribution to increased disaster risk from developed nations through pollution that provokes global warming and climatic instability. This concern about the impact of pollution on global climatic disaster occurs in a context of not knowing the precise contribution of carbon dioxide emissions to climate change, which remains a scientifically contested field. However, dependency on fossil fuels which pollute at the local level and are not in sustainable supply is one of the dilemmas of industrial development that will probably persist for decades to come. Risk creation will be likely to be transferred away from the richest nations, which may develop cleaner fuel supplies than the newly emergent economies of Asia.

The legacy of development has been the disasters it has inadvertently provoked, such as mass displacements through industrialisation, the demand for energy, economic boom and bust, erosion of livelihoods, development induced conflict, environmental degradation, and social and economic collapse that causes vulnerability to hazards. However, whilst development can play its part in increasing hazards and human vulnerability to disaster, it has also extended life expectancy and

Plate 1.6 High-rise and low-rise hazards and vulnerabilities in Dhaka, Bangladesh

Source: Author.

wellbeing, and is providing some means of protection against disaster. The limitations have been more to do with uneven development and development that is not appropriately applied to different environmental, social or economic contexts. Uneven development here is considered to be where people lack the basic standards of living, such as access to sufficient clean water, nutrition, shelter, fuel, sanitation, education and rights. Underdevelopment persists in the poorest nations, whilst rich nations progress their agendas of increased affluence, creating extreme contrasts in wellbeing. However, disaster scenarios can also have a levelling effect and raise the spectre of the inadequacy of development in reducing disaster risk. This is because disaster risks with their origins in global economic crises or macro-environmental processes can impact on all parts of the world at short notice.

Some disasters, such as Hurricane Katrina in 2005 in the southern United States (Box 1.2), demonstrated that the state had been unprepared to incur the cost of repairs to New Orleans' safeguard system, leading to it being inadequately protected. There was also a failure by FEMA to respond in an appropriate manner to this catastrophic episode. This demonstrates that rich nations, with apparently large investments in disaster preparedness, are not necessarily better protected from environmental events, though resultant loss of life is usually less.

Furthermore, mass displacements through disasters become transboundary flows of people. If we extend disaster to conflict and international security issues, the field of transboundary threats is expanded still further. In this instance the hazards can be interpreted in terms of differences in political ideology and macro-scale organisation of economies through to clashes of cultures, religious identities and secularism.

A positive view is that, as development progresses, people become less vulnerable and less likely to be in the wrong place at the wrong time without adequate means of protection. There is evidence that this is true. Thus, increased food security reduces the risk of a famine, appropriate river catchment management reduces the risk of flooding, flexible jointing of concrete structures in Japan reduces building collapse during earthquakes, and education reduces people's risk-taking behaviour, to mention but a few. Development is also the means to forewarn about disaster through scientifically evaluated risk assessment and early warning systems. These are assisted by technological developments, such as the use of remote sensing to identify vegetation loss, or of seismic modelling to try to predict the likely timing and location of earthquakes.

Box 1.2

Hurricane Katrina

Hurricane Katrina hit southeast Louisiana as a Category 3 hurricane on 29 August 2005. The most severe loss of life and property damage occurred in New Orleans, Louisiana, due to storm surge and flooding as the levee system failed at more than fifty points. Eventually 80 per cent of the city was flooded and also large tracts of neighbouring areas. The floodwaters remained for weeks. At least 1,836 people lost their lives in the actual hurricane and in the subsequent floods, making this the deadliest US hurricane since the 1928 Okeechobee Hurricane. The storm is estimated to have been responsible for US$ 81.2 billion of damage, making it the costliest environmental disaster in US history. Three years later, thousands of displaced residents of Mississippi and Louisiana were still living in trailers.

The failure of the New Orleans flood protection system prompted a congressional investigation into the Army Corps of Engineers, which had responsibility for it. Other congressional investigations were launched into the response of the federal, state and local governments, resulting in the resignation of the Federal Emergency Management Agency (FEMA) director. The National Hurricane Center and National Weather Service were widely commended for accurate forecasts.

A compendium of shortcomings in the development of protection for the residents of this area has variously been highlighted, including the following, which have been emphasised as key by Alexander (2008):

- impact scenarios were ignored and forecasts shrugged off;
- there were weaknesses in the design of the defences, and then artificial reliance on these fallible structures;
- there was an inadequate emergency plan, evaluation and management;
- the federal response was slow and inefficient;
- lessons from earlier hurricanes, such as Ivan, had not been learnt.

Source: basic facts compiled from multiple sources (see Wikipedia 2008 on Katrina) and analysis presented by Alexander (2008) at the *Dealing with Disasters* conference.

To mitigate a disaster is to develop the means to reduce its impact. Mitigation is different from avoidance, in that it assumes that a risk will start to become a reality at some point, and therefore preparedness to intervene begins to be activated. Whilst development can make at-risk people and places more resilient, catastrophic events can set back a nation by many years in development terms. The impact and recovery time has been proven to depend on what development trajectory a nation might already be on. Recovery has been rapid in Mozambique following the great flood of 2000, which is likely due to the fact that the country was on an upward long-term trend after thirty years of conflict that only ended in 1992. Examples of successful prevention measures initiated in response to previous disasters include the flood protection schemes of Bangladesh (Chowdhury 2003) and the Netherlands, or those adjusted building designs in Japan which now significantly protect or reduce the impact of earthquakes.

Future disaster reduction through development, or improved development through disaster reduction, can be considered via different strands of sustainability thinking – environmental, social and economic, as explored further in Chapter 2. Meanwhile, when development fails to prevent disasters in the economically poorer nations, a further risk is the burgeoning of a disaster or development industry that would try to correct situations based on singular lines of explanation. Thus, a disaster and development approach is one that does not assume that disasters can be dealt with by more emergency relief, faster moving response systems, economic restructuring programmes, environmental engineering or other technologies. It emphasises that disasters can be significantly reduced through social, economic or environmentally grounded early warning, conflict resolution and sustainable development initiatives. If the underlying influences that predispose people and places to disasters are addressed, there are benefits in terms of both disaster reduction and improved development.

Achieving development through disaster reduction

As is examined in more detail in Chapter 3 of this book, the cause and effect equation can also be put around the other way to explore the impact of disasters on development, indicating how disaster reduction advances development. This is the rationale that reducing disasters contributes to our achieving the MDG targets and other related relationships. The basic approach is referred to in current policy literature as mainstreaming disaster reduction into development (DFID 2005). We have seen in Table 1.1 examples of how MDGs may be compromised by disasters. Disasters either slow down or reverse development by damaging infrastructure, removing livelihoods and increasing health risks. Resources that might otherwise have been directed to poverty reduction are instead diverted into relief work. Furthermore, underlying issues of conflict resolution, strengthening community relations and civic pride may also be neglected during disaster. These are then replaced by displacement and dependency on external support. On the other hand, disasters are often accompanied by monumental examples of community cooperation and personal sacrifice, such as those witnessed in 2008 in the case of local and regionally placed citizen responders to the Burma typhoon and China earthquake. Disasters inform society and can accentuate learning for dealing with disasters. This is a theme that is revisited in Chapter 5, on learning and planning in disaster management.

The economic impacts of disasters are through loss of productivity and through slowing longer-term economic potential. However, whilst more extreme cases would be the loss of an entire crop through flooding or pests, the gradual cumulative effects from small disaster events at the very local level can also have an impact. Beyond this, however, is the case where the basic fabric of society has been decimated, leaving development in ruins and unable to reclaim a development trajectory. The desperation through ongoing conflict within a harsh environmental context in Darfur, Sudan, is a case in point. Here, when asked what priority people see for improving conditions, the reply from those most affected is to provide security. Men from the villages are likely to be killed if they go out to seek food or fuel. Women are raped. Restoration of the means to development in this version of a disaster context is a particularly complex field. However, in post conflict contexts, such as Nepal, Mozambique and other locations, it has been shown that community cooperation and coordination can return at some pace once conditions are favourable (Collins and Lucas 2002). However, in post-flood Mozambique, it was also found that wealthier people with more possessions were more affected (Brouwer and Nhassengo 2006) and therefore took longer to recover their assets. Removal of the threat of recurrent disaster in these cases can be seen to be a stimulus for development. Further details on disaster response and recovery, and the evidence for the possibility of development through disaster reduction, are returned to in Chapter 7.

The disaster and development industry

Humanitarian assistance and development aid are controlled by the world's richest nations. The share of aid budgets that provides for humanitarian relief compared to the share for development has increased significantly in recent years. A detailed record of this can be accessed through information held by the Organisation for Economic Cooperation and Development (OECD). Humanitarian assistance at the international level is channelled as bilateral aid between two nations, or as the pooled development assistance of international organisations such as the United Nations family of organisations, the European Union, the World Bank and International Monetary Fund (IMF). Official development assistance follows a route from wealthy nations, such as the G8 group (United States, United Kingdom, France, Italy, Japan, Germany, Russia, Canada), to developing countries. However, non-G8 group countries, such as the Scandinavian countries, also play a major role. Most of the industry therefore connects back to this underlying flow of finance.

However, though usually of far less financial influence, recent and internationalised trends in development policy are also shaped by some of the global institutional events that have taken place over the period from the early 1970s (Table 1.2). This is not comprehensive in identifying all of the international regime building relevant to this topic, but is suitable

Table 1.2 ***International policy events that reflect a progression of development concerns***

Date	*Event*
1972	Stockholm Conference – Sustainable development concept being hinted at.
1980	Brandt Commission on North-South issues.
1982	IUCN *World Conservation Strategy* – Greater meaning to the concept of sustainable development. 10 years after Stockholm meetings.
1987	World Commission on Environment and Development (WCED) Brundtland Report. Sustainable Development definition established. Development which meets the needs of the present without compromising the ability of future generations to meet their own needs (WCED 1987: 434). The concept is more people centred. It distinctly puts human welfare above environmental/ecological sustainability and includes the concept of social equity.
1992	United Nations Conference on Environment and Development (UNCED). Response to concerns about global environmental degradation and unsustainable patterns of development. From WCED to UNCED there was a shift away from development centred concerns toward environment centred concerns. The most likely explanation was the dominant role of the Northern countries. Numerous associated conferences and activities followed this event, some of which dealt with interrelated issues.
1993	UN Conference on Human Rights, held in Vienna.
1994	International Conference on Population and Development (ICPD), held in Cairo.
1994	World Trade Conference, held in Marrakech.
1995	Fourth World Conference on Women, held in Beijing.
1995	World Summit on Social Development, held in Copenhagen.
1996	World Food Summit, held in Rome.
1997	Kyoto Global Summit on Climate Change.
1997	Rio + 5 Follow-up to UNCED at New York.
2002	Rio + 10 Follow-up to UNCED. Johannesburg Sustainable Development Conference.
2005	World Conference on Disaster Reduction (WCDR), held in Kobe, Japan.

to demonstrate some key points about the emergence of the disaster and development consensus.

Non-governmental organisations (NGOs)

The NGOs operate internationally, nationally or as local entities. Many of the larger international organisations have undergone transitions from relief to development institutions or vice versa in much of their work, such as in the case of Oxfam, World Vision and Christian Aid. This is a mechanism that reflects a shift in demand during major catastrophes for dealing with longer-term issues of recovery and poverty reduction. It reflects a shift in thinking to dealing with the underlying causes of disaster rather than only addressing the symptoms. Other organisations, such as the Red Cross Society, Médicins Sans Frontières (MSF) or Red R, exist for specialist response to disaster, but do not engage in development per se. There are multiple roles in dealing with disasters and, whilst recognising the multiple causality of disaster, institutions often find it more effective to specialise in certain aspects of delivering development or relief. In the past, institutions have been accused of competing with each other to provide similar services and of not being well coordinated. There are examples of humanitarian assistance programmes arriving in large numbers at the scene of an incident to stake out their territory in the relief and rehabilitation process, as occurred in the Rwandan refugee crisis of 1994 at Goma, Zaire (now Democratic Republic of Congo, DRC) (Goma Epidemiology Group 1995). The problem was not a lack of goodwill, but rather a lack of good coordination. Furthermore, where systems of state governance have been weak, the tendency has been to replace the state systems, with implications for the longer-term sustainability of the actions. With the increase in their influence, the definition of an NGO has become far from clear.

The evolution of different donor and NGO development perspectives fundamentally alters the way development is administered. The legitimacy of NGOs may be ranked on the basis of their institutional status, either influenced or directly controlled by donors and governments, though they vary immensely on this account. For example, some NGOs exist primarily through funding from government grants, whilst for others most funding comes from independent supporters. Meanwhile, in some countries, local NGOs may be denied their right to operate dependent on strict licensing by the government, or simply banned as and when the government feels it opportune to do so. Some of the classification of groups identified with development and disaster work are indicated in Table 1.3.

Table 1.3 ***Broad typology of institutions identified with disaster and development strategies***

Type of organisation	*Description*	*Institutional rationale*	*Role in disaster and development work*
Emergency services	State sector primary responders.	Deal with immediate aftermath of an incident.	A part of civil contingencies and disaster preparedness plans.
Civil society	People who are informally grouped with each other through location or their means of primary subsistence.	People independently cooperating with each other towards a common goal.	Mobilises prevention and response activities as part of ordinary life.
Civil societal organisation (CSO)	Community based function, locally more representative.	Represent coordinated bottom-up strategies that include local knowledge.	Community based disaster management (CBDM) using community response groups, risk and resilience committees or similar.
Non-governmental organisation (NGO)	Has legal institutional status and usually agreements with official donors and/or recipient governments. May be national or international, with an ideological, humanitarian or religious mission or purpose.	Development or disaster reduction through projects and advocacy. Independence from the government of the country within which it was formed.	Implementing donor and government disaster and development programmes and/or emergency relief. Facilitating links between civil societal groups, funders and governments. Sensitising the public about disaster and development issues.
NGO (development)	NGO that is oriented towards human development issues.	Addresses basic and extended human needs.	Recreates livelihood security, support infrastructural development.
NGO (environment)	NGO that is oriented towards environment and conservation issues.	Addresses sustainability of the natural resource base. May be ecological conservationist or economic approaches.	Promotes intrinsic value of nature and secure natural environments. Environmental economics and natural resource management as part of disaster risk reduction and sustainable development.

Table 1.3 ***Broad typology of institutions identified with disaster and development strategies—cont'd***

Type of organisation	*Description*	*Institutional rationale*	*Role in disaster and development work*
Non-governmental humanitarian agency (NGHA)	Implements humanitarian assistance. Includes International Committee of the Red Cross, International Federation of Red Cross and Red Crescent Societies and MSF.	Saves life in emergency situations. Rationale may vary depending on the mission statement of each organisation.	Assesses emergency aid requirements and delivers to target populations during crisis. This may also be done in collaboration with or parallel to the other NGO groups.
Inter-governmental organisation (IGO)	Organisation where two or more governments represented (i.e. all of the UN agencies).	Represents international state-level dialogue and policy on issues of global concern.	Synthesises global disaster and development policy. Provides a support base to international disaster and development related strategies.
Private sector	Privately owned enterprises.	Business and enterprise for profit.	Implements strategies that improve business continuity. Ethical investment for disaster risk reduction. Makes donations to disaster and development work. Engaged in design, manufacturing or supply to the disaster and development sector.

It is important to note in Table 1.3 that the rationale of the NGO may be based on social justice, as in the case of Oxfam. Others are more centred on upholding international law, as in the case of Amnesty International. A wide group of health related institutions have been oriented by a code of conduct referred to in the health professions as the Hippocratic Oath. This is named after Hippocrates (430BC) who was a famous physician at Kos. Those who take the oath (typically medical doctors and other public health workers) are committed to assisting anyone in medical need regardless of politics, race or creed. An example would be the very successful group Médicins sans Frontières, which translates to 'Medics without Boundaries' in English. The basic vision and purpose through which an institution may have been driven, be it justice, advocacy, the Hippocratic Oath, a faith, or simply goodwill, a sense of duty or self-interest, can have some influence on where, when and with whom it will

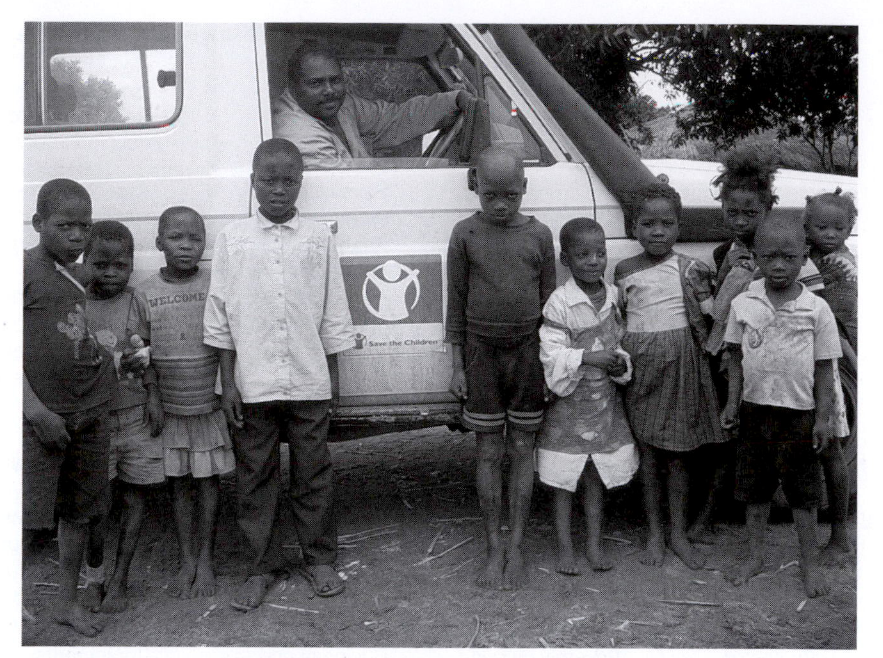

Plate 1.7 Save the Children, an international NGO that helps children in development and in disasters
Source: Author.

operate. For example, some relief institutions would not assist the Hutu refugee group, a community associated with militias that committed genocide in Rwanda in 1994, whilst others would do so. Religious belief is a motivation behind the work of some NGOs. These have their origins in, or are currently active in promoting, a faith. A long history goes back to the missionary zeal and quest to assist those in need during former centuries. There are now examples for all major religions, including organisations that operate with their religious affiliation more in the background, such as Christian Aid, CAFOD, Islamic Relief, the Agha Khan Foundation or Jewish Refugee Services. By way of contrast there are what in Western countries are termed 'evangelical organisations'. Directly associated with churches, they can sometimes combine development and relief with open and active proselytising of a faith. Beyond various motivations of NGOs, it is also fair to point out that the sector provides employment for graduates and other practitioners of disaster and development, and that for many working in it this industry is a means to a living. As such, it is reasonable to include 'self-interest' of employees as a reason for the work of the disaster and development sector.

Following the missionary tradition of some NGOs, there were links also to colonial or anti-colonial institutions. This meant that in the developing world post-independence, NGOs sprung up in response to the need for emergency relief or grew up as solidarity groups – politically and developmentally oriented. Some have straddled that divide to provide advocacy and direct practical support (Fowler 1997), such as, for example, the Fair Trade NGOs, movements such as Anti-Slavery, Anti-Apartheid, War on Want, Quaker Peace and Service, Campaign Against the Arms Trade, Jubilee 2000, World Development Movement (WDM) and a host of other worthy solidarity groups of recent years. However, with a more pronounced link between wider development policy, development programming and project funding there is a view that some mainstream NGOs now tend to directly implement donor government development ideals and policy. This raises the question of whether Northern governments' policies are simply acceptable to NGOs, or whether they just need the funding to keep going regardless of the policy of the day. Some of the dilemmas and institutional complexities of the industry can be followed up in more detail through the work of Hulme and Edwards (1997), Uphoff (1993) and Atampugre (1997).

In practice, those institutions that find themselves moving more toward the disaster and development nexus variously aim to deliver relief interventions, development options and capacity building. To justify their role they must aim both to satisfy recipients of assistance and to please their various donors. To ensure an impact NGOs are increasingly engaging in bottom-up forms of development. This comes about in response to awareness that only well-grounded locally owned initiatives are likely to progress successfully beyond the life of a project. It is a recognition of the principles of participation paradigm that expanded during the 1990s, as articulated in texts such as those of Chambers (1997, 2005). The bottom line here is that by handing over powers of decision making to communities and local individuals more effective strategies are achieved. The approach is, however, one that has been at odds with aspects of disaster management for which traditionally a structure of 'command and control' of circumstances by trained professionals is often assumed to be a priority (Neal and Phillips 1995). Nonetheless, the UNISDR, UNDP and other convinced international disaster risk reduction stakeholders are finding converts in at-risk nations to the shift from a command and control structure to a community based disaster management approach. This also allows more reliable field data on hazards, risks and vulnerabilities to be identified, and response and mitigation strategies to be articulated and initiated.

Differences in cultures of participation and representation in contrast to authority and expert knowledge have presented a dilemma in disaster prevention and response, but less so in developing areas, where there may be more opportunity for the former. With development funding, donors have since the 1980s often wanted the government sector to become more decentralised to encourage localised initiatives that are less top down. This has included encouraging projects that were thought to be more inclusive of civil society in aspects of decision making. For example, a move to more decentralised governance has been evident in the field of community based natural resource management, Ribot (2002) calculating that at least sixty countries have carried out some form of local policy development along these lines. It is also upheld as an important way forward within the framework analysis of the Millennium Ecosystem Assessment (2005), which links the ecological concerns of ecosystem services and constituents of wellbeing. However, the ideal that decentralised management of ecosystem services can support both practical conservation and economic development at the same time is producing variable accounts of success and failure (Tallis *et al.* 2008).

In disasters, the manner in which community based disaster initiatives interact with emergency relief or established emergency services is only recently attracting significant attention. It is reflected in part in the deliberations of the Hyogo Framework for Action (UNISDR 2005) (Box 1.3), which refers to a 'culture of safety and resilience at all levels' and 'effective response at all levels'. There are various examples of community based strategies in the chapters that follow, pointing to the issue of communities living with hazards (King and Cottrell 2007) and the characteristics of resilient communities (Twigg 2007a). However, in disaster management we are yet to see the full rise of the decentralised community based panacea for disaster reduction. Whilst examples of attempts to decentralise do exist, such as in the establishment of the Local Resilience Forums in the United Kingdom, the mechanism and understanding of what aspects of disaster prevention and response people at the community level can be effectively obliged to carry out constitutes a growing debate.

As NGOs increasingly become the implementers of government policy in developing world areas and are funded accordingly, they are increasingly subject to rigorous monitoring and evaluation of their actions against programme targets. In a similar way, bilateral aid to governments must demonstrate at that level that it fits programme policy and promotes good government, or face conditionality measures, whereby aid can be withheld. The idea of accountability is posed as a good thing by aid donors, but some creativity and adaptability may be at risk of being sacrificed as a result should the policies of intensive scrutiny from the top be imposed too strongly.

Table 1.3 also indicates some other categories of organisation in disaster and development work. These include civil society, which is meant to mean the less formalised grouping of lay people in any society. For these, engagement with disaster and development initiatives is through everyday life. It is possible that the potential for behavioural change at this level is one of the most powerful tools we have for disaster risk reduction. The private sector is also flagged in this list. The responsibilities here include the implementation of strategies that improve business continuity, ethical investment and donating to disaster and development work. Some private enterprises can be even more directly linked to the disaster and development agenda through being engaged in the design, manufacture or supply of technology and goods to the sector.

National Development Plans

The control by international finance of developing countries' planning increased in the latter post-independence period when international organisations such as the World Bank and IMF pushed for diminished state control. Meanwhile, NGOs characteristically tried to increase bottom-up development. The two acting together can be seen as a loss of sovereignty of the nation-state challenging the role of development and relief in terms of strengthened national governance. Detailed examples of this process are available in the case of Mozambique (Hanlon 1991, 1996). Despite broad moves towards a sustainable development discourse by most governments since the 1990s, there has only been general understanding of what it is to address. This is now joined by the disaster risk reduction agenda outlined in the next section. The issue concerns how sustainable development and disaster reduction can best be brought about and the roles of the different key players (international, regional, national, local) in achieving it.

In the current context of the MDGs this means more support for governmental Poverty Reduction Strategy Papers (PRSPs) for development assistance. These represent overarching frameworks and standards within which organizations should fit. In the case of humanitarian assistance, there is an external multi-institutional body, the Disasters Emergency Committee (DEC). This is made up of a membership of organisations that do disaster management work. In the event of a major disaster event funds are centrally managed by this Committee before disbursement to the respective organisations (see DEC website on p. 45). Other initiatives have been aimed at improving coordination on guidelines for minimum standards in disaster relief. This is the role of the Sphere Project (Chapter 7), which came about as a multi-institution initiative based on principles of good practice in humanitarian assistance, and indicates what should be delivered in terms of relief. In this instance standardisation is based more on common standards than centralisation or a top-down approach.

Introducing the disaster risk reduction agenda

With the rise of a recognised possibility of preventing disaster, rather than only responding to it, we have witnessed the emergence of risk reduction as an international aim. It is encapsulated in the UNISDR programme of activity which grew out of the Yokohama Initiative, and the UN International Decade for Natural Disaster Reduction. This laid the groundwork for many initiatives achieved during the beginning of ISDR, eventually leading to the United Nations World Conference on Disaster Reduction (WCDR) held at Kobe in 2005 (see UNISDR website on p. 44). The WCDR produced the Hyogo Framework for Action (HFA), named thus as the international talks on this topic took place in this region of Japan, on the site of and ten years after the Kobe earthquake of 1995. A summary of commitments for the period 2005–2015 are presented in Box 1.3.

The risk reduction approach is dependent on the development of accurate risk assessment. Risk assessment is in turn dependent on the application of suitable research and monitoring tools that can sense when risks are changing, in relation to what and for whom. Risk assessment therefore involves a further cycle of learning that builds on Figure 1.5. It potentially draws on many forms of quantitative and qualitative data collection, and on an ongoing process of evaluation. Accompanying this is the field of *risk governance*, which represents the way in which risk is managed or owned by different players. Risk governance is the manner in which governments, the international community, the public or private sectors, local communities, households and individuals take responsibility for managing the risk of a disaster.

Conclusion: linking disaster and development

The increasing concern about disasters can be interpreted in different ways, which has an influence on how disaster and development issues are

Box 1.3

Hyogo Framework for Action: Summary of Commitments 2005–2015

Expected outcome

The substantial reduction of disaster losses, in lives and in the social, economic and environmental assets of communities and countries.

Strategic goals

- the integration of disaster risk reduction into sustainable development policies and planning;
- development and strengthening of institutions, mechanisms and capacities to build resilience to hazards;
- the systematic incorporation of risk reduction approaches into the implementation of emergency preparedness, response and recovery.

Priorities for action

- ensure that disaster risk reduction is both a national and a local priority with a strong institutional basis for implementation;
- identify, assess and monitor disaster risks and enhance early warning;
- use knowledge, innovation and education to build a culture of safety and resilience at all levels;
- reduce the underlying risk factors;
- strengthen disaster preparedness for effective response at all levels.

Source: World Conference of Disaster Reduction (WCDR) (2005).

addressed as interlinked and on the nature of response that may be expected from different institutions. Other than the fact that development can reduce disaster, and is also the response we turn to in the wake of disaster, the link between the two requires close examination. Some of the theoretical underpinning for dealing with disasters suggests there are crucial possibilities for choices in the interests of disaster reduction. No disaster is therefore inevitable.

Discussion questions

1 Which critical events are readily referred to as disasters? What do we mean by a disaster in terms of development?
2 How might different disciplines contribute to reducing vulnerability?

3 In what way is it not inevitable that there are major disasters?
4 How do we address disaster reduction as a cyclical process of prevention and response?
5 What might be the possibility of achieving Millennium Development Goals through disaster reduction?
6 Are NGOs the best route for the delivery of development and relief? What are the alternatives?

Further reading

Short lists of recommended items cited in this book are provided after each chapter.

Alexander, D. (2000) *Confronting Catastrophe*, Harpenden: Terra.
DFID (2005) *Disaster Risk Reduction: a development concern*, London: DFID.
DFID (2006) *Reducing the Risk of Disasters: helping to achieve sustainable poverty reduction in a vulnerable world: a DFID policy paper*, London: DFID.
McEntire, D.A. (2004) 'Development, Disasters and Vulnerability: a discussion of divergent theories and the need for their integration', *Disaster Prevention and Management*, 13:3, pp. 193–198.
Quarantelli, E.L. (ed.) (1998) *What Is a Disaster? Perspectives on the question*, London: Routledge.
Tudor Rose (2005) *Know Risk*, Geneva: United Nations.
Twigg, J. (2007) 'Disaster Reduction Terminology: a common sense approach', *Humanitarian Exchange* 38, pp. 2–5.
UNDP (2004) *Reducing Disaster Risk: a challenge for development*, Geneva: Bureaux for Crisis Prevention and Recovery. Available from: http://www.undp.org/bcpr/disred/english/publications/rdr.htm.
Wisner, B., Blaikie, P., Cannon, T. and Davis, I. (2004) *At Risk: natural hazards, people's vulnerability and disasters*, second edition, London: Routledge.
World Disasters Report (annually since 2002).

Useful websites

http://www.em-dat.net/index.htm: EM-DAT, the International Disaster Database. Centre for Research on the Epidemiology of Disasters (CRED), Ecole de Santé Publique, Université Catholique de Louvain, Brussels, Belgium.
www.unisdr.org: United Nations International Strategy for Disaster Reduction (UNISDR).
www.preventionweb.net: this is an information service of the UNISDR.
www.ifrc.org: International Federation of Red Cross and Red Crescent Societies.
www.proventionconsortium.org: the ProVention Secretariat is hosted at the IFRC, Geneva, and focuses on disaster risk reduction.
www.reliefweb.org: for updates on disaster situations around the world.
www.alertnet.org: this site is run by the Reuters Foundation and provides alerts on emergencies aimed, in particular, at humanitarian agencies.

www.dec.org.uk: Disasters Emergency Committee (DEC) coordinates emergency aid.
www.eldis.org: portal for development related information run by the Institute of Development Studies, University of Sussex.
www.odihpn.org: Overseas Development Institute (ODI) Humanitarian Practice Network.

Note on journal articles

Articles relevant to this field appear in a wide range of journals. The following are recommended for an online search. Web addresses change from time to time, so start by just searching for the journal title to locate it.

Most directly focused on disaster management, including development aspects:

- *The Australian Journal of Emergency Management*
- *Disaster Prevention and Management*
- *Disasters: The Journal of Disaster Studies, Policy and Management*
- *Environmental Hazards: Global Environmental Change B*
- *Humanitarian Exchange*, Humanitarian Practice Network, Overseas Development Institute, UK
- *Journal of Contingencies and Crisis Management*
- *Journal of Emergency Management*
- *Journal of Mass Emergencies and Disasters*

Addressing disaster and development topics indirectly, from time to time including papers on disasters:

- *Bulletin of the World Health Organization*
- *Development Policy Review*
- *Disaster Management and Response*
- *Environment and Urbanisation*
- *Forced Migration Review*
- *GeoJournal*
- *Global Environmental Change*
- *International Journal of Environmental Studies*
- *Journal of Refugee Studies*
- *The Lancet*
- *Land Degradation and Development*
- *Progress in Human Geography*
- *Social Science and Medicine*
- *Water Management*
- *World Development*

2 Viewing disasters from perspectives of development

Summary

- Interpretations of disasters in terms of development range from fatalistic views of unsustainable development to more optimistic versions of the human capacity to cope, adapt and prosper.
- Excessive disruption of the natural resource base increases human vulnerability and environmental hazards, but these are also mitigated by technological development.
- Poverty and disaster vulnerability are multidimensional conditions demanding integrated reduction strategies.
- Globalisation and urbanisation intensify disaster risks.
- Disasters, human insecurity and complex emergencies can be addressed through a comprehensive approach to sustainable development.

Introduction

This chapter outlines debates that concern the nature of development induced disaster, with reference to some well-established ideas about human survival and wellbeing. It considers how environmental degradation, poverty and underdevelopment are multidimensional conditions that underlie many of the world's major disasters. There are no simple explanations as to why the quest for development should induce human disasters. People generally desire improvements in wellbeing as a human instinct and right, minimal standards of living alone remaining the aspiration of few, if any, societies. However, the drive and options for material improvement have varied in time and from place to place. In disaster analysis, a desire for increase has often been recognised as being at the expense of the stability of the world's natural resource base. A pessimistic conclusion is that the promotion of short-term improvements will always outweigh a sense of longer-term preservation of life. This in turn drives the earth and its people towards disaster.

However, the optimistic view is that long-term development demands can be mediated by what we have termed appropriate technologies and sustainable development. The optimist believes that common sense, conservation and the development of better technologies can sufficiently mitigate world disaster scenarios. Underlying either version of events is the sustainability debate addressed in this book and the series of which this book is a part. Ideas about sustainability are based on the importance of how we can act now to maintain the rights of future generations to choose to avoid future disaster scenarios. The chapter therefore is concerned with the varying interpretations of the impact of development on disasters. This includes sections on development and environmental determinism, poverty, economy, urbanisation and complex emergency. However, given the centrality of sustainable development to avoiding disasters, we pause briefly at the start of this chapter to consider the core sustainability perspectives and their broader relation to thinking on disasters.

Sustainability and disasters

Table 2.1 indicates three accepted sustainable development emphases. It is developed from a framework approach implied by many development thinkers, particularly following Brundtland's version of sustainable development in 1987 (WCED 1987). The first three rows of the table are an adaptation of a larger framework produced by Hatzius (1996). Some parts of that are summarised here, with the addition of the relationship of sustainability to disasters across three variants of sustainable development thinking: ecological, social and economic. In terms of ecological sustainability (first column) with a purpose of ecological viability and a policy rationale to protect nature and the environment, disasters concern impacts on people resulting from catastrophic changes in the environment. This is what in the disasters field of study has traditionally been termed a 'natural hazards approach'. It includes human induced environmental disasters, particularly where environment is

Table 2.1 ***Viewing disasters from different sustainability perspectives***

	Ecological sustainability	*Sustainable development*	*Sustainable growth*
Purpose	Ecological viability	Social efficiency, justice	Economic efficiency, sustainable production and reproduction
Policy rationale	Protect nature, educate people, equilibrium, holism, co-evolutionary ideas	Empower people, build community, develop institutions and livelihoods	Develop markets and internalise externalities
Relationship of sustainability issues to disasters	Environmental hazards and people, environmental change	Vulnerability, human security and multiple dimensions of poverty	Institutional security, infrastructure, economic policy

extended to mean its industrialised aspects. For sustainable development (second column) with a more social focus, there is an emphasis on putting people, community and livelihoods first, disasters in this way being interpreted more specifically in terms of human security and poverty. The third column represents the case where sustainable development is considered more in terms of sustainable economic growth, particularly market based production. In this instance, disasters become more oriented to institutional security, infrastructure and economic policies.

In reality, we might consider the type of sustainable development required for disaster reduction to be about all of these components. Indeed, the earlier descriptions of sustainable development proposed by Brundtland (WCED 1987) advocate an integrated strategy. Ultimately, a multiple aspect sustainable development such as this may be considered a form of integrated disaster reduction. This is because integrated disaster reduction involves engaging the ecological, social and economic aspects of sustainable development. Integrated disaster reduction is a theme central to current disaster management thinking and one returned to at several points later in this text.

Malthusianism and disaster

A fatalistic, determinist and structural view of disaster is evident in the ideas of Thomas Malthus in the 1790s and early 1800s. He proposed that the pressure of a rapidly expanding human population would lead to too much demand for the earth's resources and ultimately to disaster. His main argument was that food supply increases gradually from year to year, but that population increase is exponential due to people's desire to continuously have children beyond the number needed to replace themselves. Public perception that there are too many people in the world to feed has been a recurrent idea in the generations that have followed Malthus's era. Regarding absolute numbers of people in the world the Malthus prediction of expansion has not been far from the reality. Whilst in 1950 there were about 2.5 billion people in the world, by 2000 there were 6 billion, a rise that exceeds that in the preceding 4 million years (United Nations 2003). However, catastrophic consequences in terms of food supply are less obvious.

Despite the ongoing presence of famine, we know from WHO and Food and Agriculture Organisation (FAO) data on food consumption (kcal per capita per day) that availability of calories per capita from the mid-1960s to the late 1990s increased globally by approximately 450 kcal per capita per day and by over 600 kcal per capita per day in developing countries (WHO 2008b). This means that the average availability rose from at least 2,258 kcal in the mid-1960s to 2,803 kcal at the end of the century, leading to predictions of 2,940 kcal for the year 2015 and 3,050 kcal for the year 2030 (FAOSTAT 2008). This far outstrips basic food energy requirements of about 2,100 kcal per capita. However, the change was not equal across regions. The per capita supply of calories remained almost stagnant in sub-Saharan Africa, which has experienced calorie availability at levels below the minimum level required, and recently fell in the countries in economic transition. In contrast, the per capita supply of food energy had been rising dramatically in East Asia (by almost 1,000 kcal per capita per day, mainly in China) and in the Near East/North Africa region (by over 700 kcal per capita per day). However, despite this overall counter-Malthusian trend in terms of food supply, it is cereal production that has more recently been looking unstable. FAOSTAT (2008) records that in 2006 both world agricultural production and food production rose by less than 1 per cent, with consequent estimated per capita agricultural food production falling by about 0.2 per cent. Some view the outlook very gloomily, predicting a worsening of food security over the coming years as efforts to increase quantities of production compete with industrialisation. A theory advanced by Brown (2005) in relation to staple grain diets and through noting the case of Japan is that 'if countries are already densely populated when they begin to industrialize rapidly, three things happen in quick succession to make them heavily dependent on grain imports: grain consumption climbs as incomes rise, grainland area shrinks, and grain production falls. The rapid industrialization that drives up demand simultaneously shrinks the cropland area. The inevitable result is that grain imports soar'

(Brown 2005: 11). As more countries start to fit this pattern there emerges an overall shortage of producer regions in relation to importing regions.

Reasons for the overall lack of a correlation between population increase and food supply problems over the second half of the twentieth century were that new technologies and methods address overall food needs (Boserup 1965). Furthermore, Sen (1981) provided an emphasis on famine as being essentially about the command people have over food and other resources, otherwise described as 'entitlements'. The issue of access to resources, entitlements and ultimately rights underpins the wider field of a rights based approach to development. From here it is only a short step in our understanding to also be able to accept the importance of a rights based approach to disaster reduction. Nonetheless, beyond the issue of food supply, the earlier ideas promulgated by Malthus still in part resonate in the literature of more modern times. Ehrlich (1962) wrote the text *The Population Bomb* to this effect and, following its regular citation by environmentally concerned people through the decades that followed, repeated the same sentiments with the publication of *The Population Explosion* (Ehrlich and Ehrlich 1990). The durability of these ideas has been reinforced by the widespread evidence of environmental degradation and how, unchecked, this paves the way for human disaster through loss of a sustainable resource base.

Malthusian ideas about expansion of demand for resources also still have resonance inasmuch as overpopulation can cause pollution beyond the environment's capacity to absorb it. If not population, then industrialisation, which in some cases has slowed increases in the birth rate, but brings its own chemical pollution. Reflecting the concerns of the day, Rachel Carson's book '*Silent Spring*' (Carson 1962) brought to the attention of the public in the US and beyond how industrial pollution was, in effect, killing nature. This hit home at a time considered to be the beginning of the current environmental movement and environmentalism in Western culture. It is also now the concern raised by the destruction of ozone or increase in greenhouse gases, where development brings unremitting release of atmospheric pollutants. Whereas this was previously the concern of the so-called developed nations, they are joined now by newly industrialising countries or those undergoing rapid economic change, such as India and China.

In 1968 Garrett Hardin came up with a theory that, from a neo-Malthusian angle, tries to explain why humankind does this to itself. This theory, called 'The Tragedy of the Commons' (Hardin 1968), has been used not only to explain the mindset of people who destroy the environment they need for survival, but also to address questions about the management or governance that is needed to address that condition. It represented a further model against which the real world might be observed, but is revealing in terms of ideas about disaster and development, and is therefore explored a little further here.

The use of the tragedy of the commons theory as an explanation for development induced disaster

The tragedy of the commons idea states that the individual gain to each user from overusing the commons will always outweigh the individual losses that he or she has to bear due to its resulting degradation. Hardin (1968) was referring to commons as resources for which there may be notionally no formal ownership, such as rangelands in East Africa. With pasture open to all, each herdsman will try to keep as many cattle as possible. Wars, poaching and disease might keep the numbers of people and cattle below the carrying capacity of the land. However, as a rational individual, each herdsman seeks to maximize his gain. He calculates the utility to himself of adding an extra animal. As an individual, his gain is greater than what he loses as a result of there being additional pressure on the land for grazing. He concludes each time that he gains by adding one more head of cattle. In the short term this is correct, but the same conclusion is reached by all the other rational thinking herdsmen using the common pasture, leading to overgrazing.

Each herdsman is locked into a mindset or system that compels him to increase his herd size, driven by a personal sense of development. It is therefore the freedom in the use of the common resource that adds pressure on the environment to the point of disaster. Though individuals may benefit, through their denial that the earth's systems will be dangerously degraded, the society of which they are a part will suffer. Education can counteract these 'natural' tendencies, but education or the lessons learnt from this experience can diminish over time. Different tensions acting on rational behaviour concerning decision making and governance of the environment might therefore be summarised along the lines of Figure 2.1.

The idea of a tragedy of the commons has been likened to parallel examples where trust and cooperative behaviour between 'rational' individuals are required. Bryant and Bailey (1997) use the example of the Prisoner's Dilemma. This uses the hypothetical example of two men

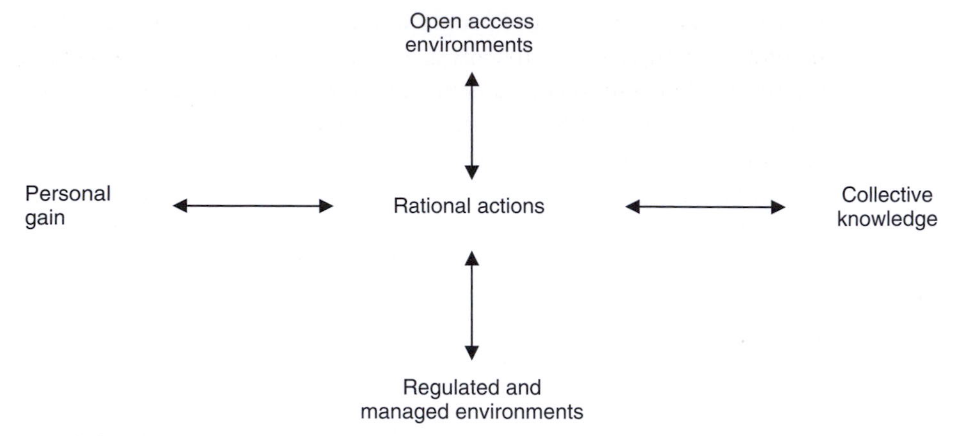

Figure 2.1 Rational influences on individual and group environmental behaviour

arrested by the police for a minor offence, but who are each suspected of being responsible for a serious crime. The police interview the two suspects separately and, to acquire conclusive evidence, present each man with a deal: squeal on your accomplice and if he is convicted of the major crime go free, or stay silent and go to jail by being convicted of the lesser offence. The collective interest of both men is to stay silent, since that option provides the lowest overall sentence outcome. However, it is in each man's individual interest to squeal, provided that the other man does not follow suit, because that way he is able to escape any punishment. Both men separately reach that conclusion and squeal accordingly, with the result that they are both then convicted of the major crime. This result is the worst possible collective solution and illustrates the Prisoner's Dilemma – namely, that in the absence of trust (or a state to enforce 'trust') between individuals, it is inevitable that people will act individually, but with an outcome that is sub-optimal from the point of view of society as a whole. It is possible to extend this principle to cooperation over a wide range of common property environments, from conservation of physical environmental resources to knowledge. The concern here is that the takers and plunderers acting on individual rationales threaten us all.

In particular, the tragedy of the commons logic is applied to environmental pollution and has particular resonance when we consider why some people choose to recycle their waste whilst others do not. The net level of degradation to the environment results from human inputs such as sewage, chemicals, radioactive waste, air pollution and heat discharged into common resources to the gain of individual polluters or industries. The 'rational' person or enterprise acknowledges its individual, company or corporate gain through this process, but tends to remain conveniently ignorant of the implications for human development as a whole. For example, plantation owners in Indonesia feel compelled to burn forests to increase crop size, but cause smoke pollution all over Southeast Asia, with public health consequences when they do so. Industrialised countries, most notably the USA, Europe and more recently India and China, belch high dosages of pollutants into the atmosphere. This causes net increases in environmental instability through greenhouse gas induced climate change and the associated risks are collectively acknowledged (Helmer and Hilhorst 2006; Schipper and Pelling 2006; O'Brien *et al*. 2006; van Aalst 2006). Hardin tried to point out that we are only locked into a system of 'fouling our own nest' so long as we behave as independent, rational free enterprisers. The question has therefore been more about what global system of environmental monitoring and control is needed to police this form of disaster risk.

However, there is a limit to Hardin's rationale for explaining the whole scenario of environmental degradation. This is because there is evidence, particularly at the local level, that in the face of disaster human cooperation over conservation may prevail. For example, in parts of the UK where there is a coastal erosion threat, people tend to adhere to using recommended pathways through managed marram, lime grass and sand sedge conservation schemes that promote sand dune stability and coastal zone protection. The point is that even where an area is considered as being for the use of the public, a common interest and intelligence in protecting it may prevail. The reason why the ultimate destruction of nature by humankind is for many too fatalistic a view is that it is an egocentric perspective of the world only. That is to say that it functions on the basis of rational self-interest to development. An egocentric view, regarding everything only in relation to oneself, is where the individual operates on the basis that they personally will not be around later to suffer the consequences of their actions when it turns to disaster. This needs to be contrasted with what has been termed an 'ecocentric' view (after O'Riordan 1976), which makes the ecosystem, and a deeper understanding of the importance of ecology, the prime focus for our objective or rational understanding of environments.

With this approach our purpose in the environment and options of long-term sustainable conservation are included. In its greenest interpretation, this is along the lines of what has been referred to as a deep ecology approach (Pepper 1984, 1996). An alternative to ecocentrism, but similarly offering a less fatalistic option, is the technocentric perspective,

Plate 2.1 Students learning about forest management in Scotland
Source: Author.

Plate 2.2 A livelihood from woodland clearance and charcoal production in Africa
Source: Author.

where it is maintained that humanity survives and then prospers through further technological development, be that grand scale or intermediate. Unlike ecocentric views that seek to maintain balance, technocentric perspectives endorse development and change as part of human coping, a way of staying prepared and ahead of the demands of a hazardous world.

It is important to note that the metaphor of the tragedy of the commons as a planet-wide situation can be used to come to different political conclusions. It could be a socialist condemnation of individual or corporate private ownership, market freedom and self-interested economic rationality. Meanwhile, from a market economy point of view it could be seen as supportive of private ownership of everything so that liability can be identified for decisions made. It would avoid the potentially damaging effects of homogeneous ownership of resources such as that previously imposed by the communist bloc countries. In this context, a policy and motivation vacuum in large state sectors resulted in neglect of local environmental quality, leading to levels of pollution which were often as great as or in excess of those produced by capitalism. But as multilateral corporations now own increasingly more of the world's resources, globalised private ownership is of current concern to maintaining environmental and socio-economic sustainability. Regardless of the particular macro-political alignment, this is an interpretation of increasing disaster risk associated with authoritarian and monolithic approaches to development.

The implication of the above assessment is that responses to global disaster threats are based on a similar mindset. Hence, in the 1970s Malthusian thinking lay behind an expanded call for birth control. The analogy of the lifeboat was used, whereby with 10 people in a lifeboat, adrift in a sea full of drowning people and with supplies only for 10, pulling someone into the boat and sharing supplies between 11 would mean they all starve (Hardin 1974). Through this line of thinking a lay view is that there might be some sort of an argument against giving food aid to the Third World, for the threat of famine would be a self-regulating influence on population and a threat it was felt best left for nature to resolve. Conversely, international assistance in the era from the 1960s onwards might only justifiably be given to those countries pursuing a vigorous population control programme. In China, the one child policy was imposed by its own government, whilst in India forced sterilisations took place. The view that a common freedom, in this case to breed, would bring about a catastrophe was widespread. In more recent years, the fear of too much pressure on our common future has been analogous to wider ideas of global policy inertia with regard to global warming, ozone depletion and ocean pollution. This extends to arguments for and against free trade or trade protectionism and regarding the property rights of economically underrepresented groups. Though there are some cumbersome agreements, such as the more than sixty agreements on goods, services and intellectual property rights through the World Trade Organisation (WTO), the issue of how to bring about common responsibilities and viable common futures does not go away.

The 'neo-Malthusian' school of thought accepts the need for custodians of common assets to be included in law making and for legislation to be tailored to different realities of individual areas. A fuller more puritanical version recognises the need for legislators to be monitored and to find ways of keeping those in control (the gatekeepers) honest and accountable. It tends to accept that food supply is less at risk where food-producing land is mainly locally owned. However, air and water are not fenced, such that averting degradation of these resources requires a more challenging approach to those who degrade them. Making it more expensive for a polluter to emit waste into the environment than to treat waste is one tactic. This purely economic rationale, however, creates the scenario of polluters being able to pollute to a threshold, encouraging a low-level equilibrium on environmental quality. More recently, we have seen the rise of the principle of 'offsetting', whereby it is possible to pay off one's impact on pollution by investing in environmental schemes elsewhere.

The useful part of analysing theories of development and environmental degradation is in the realisation of why they might not actually fit reality. For example, there is a strong view that the tragedy of the commons does not reflect reality as the real issue of environmental conservation is the 'tragedy of enclosure', whereby privatisation or nationalisation of a resource prevents open access usage by lay people (*Ecologist* 1992; 1993; 1). Interpretations of humanity along the lines of the tragedy of the commons oversimplify matters because they make too many assumptions about people's behaviour, not recognising people who would claim to respond to rationales and moralities other than economic ones. As such, facilitating and rewarding people for making the right individual choices becomes more important than creating more and more environmental legislation for preserving the planet that is not adhered to. People care about the future welfare of their children, are influenced by religion, ecological ethics, or have a gut feeling about the need to avoid inherently risky behaviour. The evidence can be seen in the way communities have banded together throughout the ages to respond to impending crisis without there being immediately apparent benefits to themselves.

Though scientific certainty may be politically manipulated, we may successfully judge a situation through awareness of an overall context. The view of the world along the lines of what Hardin was suggesting arguably reflects a view of people as ignorant of effective local management regimes that may be applied to shared resources. For example, ancient Israel was operating its production system on the basis of seven by seven years of land ownership, with the fiftieth year a 'Jubilee' year. In the Jubilee year the land was returned to God, and then the system would start all over again. Arguably, where long-term control and ownership is removed, environments have more opportunity to be diverse or to self-regulate.

Viewed in this way there are in reality few common resources that are actually open access, but rather there are common property regimes. In the case of the latter, authority over the use of forests, water and land rests with anything ranging from a national authority to a local community or individual. Where there is a democratic system of governance, this management context can be made sustainable should public awareness be sufficiently oriented to disaster risk reduction. In this scenario, people must negotiate with each other, but profit is not the only social value operating. Far from being a 'free for all', use of the resource base can be closely regulated through communal rules and practices. Examples of community based organisation along more sustainable lines are widespread. For example, the Wabigoon Lake Ojibway Nation of Ontario, Canada, harvest wild rice regulated through community meetings. There are carefully managed common grazing or broad leafed tree planting schemes in the deforested Highlands of Scotland, and there is often cooperation over open well water systems in flood and drought prone regions of Africa (Collins 1998).

If community access to and control over resources is a step towards safety and security, then the logic might be that enclosure of resources means that local control and security is potentially lost. This is what occurred in England with a series of Acts of Parliament in the United Kingdom by which open fields and common land on which people once had rights to graze animals were lost. Enclosures occurred in England between the twelfth and fourteenth century, and again in the eighteenth and nineteenth centuries, and then with the General Enclosure Act of 1845, which allowed for Commissioners to enclose land without submitting a request to Parliament. Twenty-one per cent of the total land area of England is thought to have been enclosed through these acts. This had the effect of encouraging many English country-dwellers to move to urban areas for wage labour, interpreted in Marxist terms as forcing people into the sphere of capitalism. As urban areas were disease ridden, it forced them into areas of higher health risk.

It can be argued that as the environment varies greatly from one locale to another, its enclosure by one interest group (such as a company, private individual, local authority or international organisation) is exclusionary and therefore hazardous. Diversity of usage and the securities that the land may offer are lost along with it. If permitted, communities adapt their own checks and balances in the use of their local environment. When a sense of ownership is lost, these adjustments cease to occur, thus creating an increased risk of unsustainable use of resources.

As we shall see in subsequent chapters, community self-organisation and empowerment are vital to effective localised management of disasters. They create the rationale behind community based emergency preparedness schemes in several parts of the world, a topic returned to in chapters 5 and 6. A disaster and development approach shows that local management techniques are crucial as there are few 'natural' environments where people are not living nearby. People can therefore be successfully locked into a management cycle for local environmental quality, managing any associated hazards and risks. The issue here is to identify what type of enclosure or governance system best allows people to conserve an area and reduce disaster. One hard learnt lesson has been the growing awareness that to avoid an increase in disaster vulnerability, it is a priority to also meet the livelihood requirements of the people who live in the hazard zone. This has been identified by Wisner *et al.* as a risk-reduction objective to 'reduce risks by improving livelihood opportunities' (2004: 351).

There have been several attempts more regularly documented on the principle of using conservation enclosure and livelihood benefits to do this. One example, that of the Communal Areas Management Programme for Indigenous Resources Project (CAMPFIRE) in Zimbabwe, illustrated the point that the transfer of benefits can (theoretically) increase local investments. In this instance high value wildlife such as lions become a common resource in that the people are allocated a percentage of the payment made by the hunters who hunt the animal. Local people protect wildlife from poachers and do not hunt it themselves because it becomes a valued resource with a financial value for the community nearby, which could include funding toward a school or clinic. Whilst this is a strong idea theoretically, and one with applications beyond wildlife, Balint and Mashinya (2008) have nonetheless found that in practice there have been many shortcomings in implementing such an approach in the real world contexts of Zimbabwe.

If a similar approach was successfully applied to hardwood trees in the tropics, which may prove to be more manageable than the case of gaining local returns from wildlife, we would probably witness less destruction of that resource by illegal loggers, more controlled burning by villagers and some poverty reduction. The value of one hardwood tree is sufficient to raise the living standards of an entire village, lifting them out of poverty, but currently the wealth generated rarely remains near to the community from which it was extracted. It is a shame that more has not been made of the CAMPFIRE approach to resources beyond game animals, despite there having been some limitations in this model.

One view on the theme of development and enclosure which touches on where some of the underlying risks may lie is succinctly put in the editorial of the *Ecologist* of 1992, which stated that:

> The modern nation-state has been built only by stripping power and control from commons regimes and creating structures of governance from which the great mass of humanity (particularly women) are excluded. Likewise, the market economy has expanded primarily by enabling state and commercial interests to gain control of territory that has traditionally been used and cherished by others, and by transforming that territory - together with the people themselves - into expendable 'resources' for exploitation. By enclosing forests, the state and private enterprise have torn them out of fabrics of peasant subsistence; by providing local leaders with an outside power base, unaccountable to local people, they have undermined village checks and balances; by stimulating demand for cash goods, they have impelled villagers to seek an ever wider range of things to sell.
>
> (*Ecologist* 1992: 131, no individual author)

Such sentiments have underlain a more activist approach to saving the planet, but how in reality this might be implemented in the context of the current world system is at times less clear.

Limits to growth, and adaptation

Much of the concern with environmental degradation is encapsulated in the 'limits to growth' thesis based on ideas developed out of the work of Schumacher (1973) and also famously by Meadows *et al.* (1972) using a neo-Malthusian perspective. This was later revisited as a further work called *Beyond the Limits* (Meadows *et al.* 1992). The idea of there being limits to growth is based on 'carrying capacity'. Whilst cautionary on the chances of human survival with ensuing environmental degradation, limits to growth recognises human innovation and options for moderating development. Appropriate development acts as a counterforce to disaster, in that adverse development impacts can be moderated. Development can be brought under control, whilst people and societies are adaptable to change through limits on development. Individuals, communities, economies and the environment also have adaptive capacity that helps them cope with changes in hazards and risks. The thresholds in terms of limits to growth are therefore changeable in relation to how much people are prepared or able to adapt.

With the emphasis on human adaptation, potentially negative feedback from development can be seen as more a question of adjustments in the relationship between people and potential disasters. In this case, disasters such as flood, drought and disease epidemic can be adapted to by improving people's ability to coping with crises or by increasing resilience to environmental change. The idea that institutions and people can adapt to changing disaster risks is a focal point of the agenda associated with current concerns about disaster and development implications of climate change. An entire edition of *Disasters* (2006) is dedicated to affirming this point, also analysing potential ways forward from an understanding of the compounded nature of climate associated risks (Helmer and Hilhorst 2006). However, focusing on human adaptive capacity might arguably be an admission of failure to force polluters to remove their pollutants and more radically reverse adverse development trajectories, should that be the underlying cause of the climate crisis.

There is often a gap between those responsible for making adaptation necessary and those having to adapt. Polluters in industrialised nations do not live in flood prone areas of the tropics where survival chances are reduced by human induced climate change. Adaptation, however, is arguably a forced condition for those living among disaster hazards and already surviving the consequences of development beyond the limits to growth. The need for adaptation to climate change is an outcome that has been imposed on the poor rather than a natural or voluntary adaptation. Should the polluter be responsible for this increase in adversity, it can increasingly be argued that they should pay for the consequences and undergo fundamental development reforms.

Adaptation is a necessary concept both in the contexts of problems brought about by overconsumption of resources, such as fossil fuels, and problems of underdevelopment, where, for example, infrastructures are insufficient, and where ability to trade is weak. The ethical problem is that forced adaptation is required when economic development in some regions of the world causes underdevelopment in others. Thus, the global economy already demands daily adaptation of the poor to cope with the demands of the market. New technologies, crop varieties or more productive machinery demand adaptation of the means of production and of entire cultures to catch up or contend with development. Conflicts also demand adaptation, the ultimate response of which is coping with displacement and forced migration to safer or more productive areas. The topic of displacement and disasters is returned to in more detail in Chapter 3. Further examples of adaptation to disaster scenarios involve changes in human behaviour, such as in sexual practices in view of the risk of HIV/AIDS. We see that adaptation is rarely self-selected, but rather forced by necessity once the consequences of non-adaptation are more fully realised. Adaptation is a political and moral issue, as well as being about an individual's capacity to change in response to a disaster risk. In the contexts of disaster risk reduction an individual person's ability to adapt to a more protective lifestyle will invariable strengthen the resilience of the entire group of which they are a part.

Economic development and disaster

An economic growth dominated perspective of development and disaster emphasises either too little economic activity or unregulated over-intensive economic activity. It has a long history. The perspectives referred to here are derived from the classical and neo-liberal development theories. They are outlined in some detail in Willis (2005), in this series of books. The ideas of both Adam Smith (1723–1790) and of David Ricardo (1772–1823) were based on the idea of freeing up economies to stimulate development. This was a reaction to mercantilism that was all about protecting one's trading links. The idea of the free market has never left the world since those days of free plunder, some of its most vociferous influence being in the drive for free trade amongst Western countries during the period from the Second World War until the present. Under a free trade and essentially capitalist system every nation is alleged to benefit through the opportunity to development a 'comparative advantage' in what it is able to produce best. However, the limitation here has been the assumption of a level playing field upon which to trade freely and engage in economic development. Instead of reducing risks of economic decline, in the context of world power brokers and corruption, it has reduced the risks of economic disaster in some regions, whilst increasing them in others. Thus, during the era of European colonialism and industrialisation, economic development was accompanied by a steady increase in life expectancy through a decline in mortality and improvements in diet and medicine. However, as fertility remained high, large increases in population resulted.

Economic development leading to both improved health care intervention and more prosperous living conditions eventually brought down birth rates, stabilising the economically driven population boom. In recent times, some parts of Europe, particularly Scandinavia and parts of Eastern and Southern Europe, have witnessed a change that may have happened too quickly. Relative improvements in economic prosperity and consequent changes in society have to some extent led to a crisis referred to as 'population bust'. In this instance, potential crisis is related to there being too few economically active (tax paying) members of a population to generate wealth to support a growing percentage of elderly dependants. Decreases in population represent a subsequent step in the demographic transition model, distinctly in contrast to Malthusian views of disaster, but nonetheless with its own impact. Despite some stabilisation in population globally, as an indicator of development, some areas have maintained a 'demographic trap' over the last few decades, with persistently high fertility. In areas of sub-Saharan Africa, where both poverty and HIV/AIDS have decimated entire communities, and where HIV/AIDS represents the greatest current disaster category, any population crisis is more in terms of population structure than changes in overall numbers of people. With too few economically active people in the 16–35 age group, the burden on the very young and the very old is increased.

Successful economies bring stability for a time, whilst uncertain economic climates create instability. This was learnt in the Wall Street Crash of 1929 and the Great Depression of the 1930s. Economic downturns also provide the context for global upheavals and conflicts. Almost all of the larger conflicts of the world can be traced to a significant extent to combinations of poverty, trade tensions or accompanying imposition of ideologies offering variously to bring about stability. The failure of some market approaches was realised at various stages in history. Amongst these, and as an alternative to classical economics of earlier years, was the view of economic development as being more about an organic process of development. The most prominent economist associated with ideas along these lines was John Maynard Keynes, who in 1936 was already exploring how good investments could address issues of employment beyond public works projects (Keynes 1936). He advocated investment in the private sector to create jobs, which would in turn increase spending, indirectly creating more jobs and greater demand for goods. This is referred to as the multiplier effect and lies within the logic of more globalised economic policies that took off following the Second World War.

The post-war reconstruction period was a turning point in world history, not least through the establishment of global institutions charged with maintaining the peace. Herein lies the origins of many of the current development and humanitarian programmes. The predominant view was one of stability and prosperity through economic development and regulation. The agenda, which was very much driven by a Western view of development, was encapsulated by the Bretton Woods Conference in 1944 in the USA, which created three world institutions. These were the World Bank, the IMF and the General Agreement on Tariffs and Trade (GATT). Although it had some links to these, the wider United Nations programme developed differently and, consequently, with a much broader development rationale. The other major economic alternative in the powerful parts of the world remained the ideologies of the communist bloc, the development ideas of which varied but were in one form or another based on Marxist thinking. By way of definition of what the Western nations were pursuing, frequent reference has been made to the work of Walt Rostow (1960), who wrote the influential book *The Stages of Economic Growth: a non-communist manifesto*. Rostow suggested a linear staged approach to development, whereby primitive societies pursue development through modernisation towards an age of high consumption, much in the way that some developed nations have done.

This vision of development has been repeatedly challenged as not reflecting the environmental and social concerns of mass consumption and development. It assumes that cultures respond to ideas in similar ways, and consequently that there are linear globalised notions of development (Sachs 1993). More recent concerns are that should all of the world develop along these lines we would move even more rapidly beyond sustainable limits. Sachs questions the very notion of development itself as a forward moving process in his essay on the 'archaeology of the development idea' (Sachs 1999: 3). However, the model of economic development will always hold appeal where there is an aspiration to share in the developments which have hitherto been more the preserve of the developed nations. Modern-day South Asia is an economic boom area in terms of new cars, computers and other luxury items, all of which are increasing pollution hazards in the same way as occurred earlier in Europe and North America. Meanwhile, it is clear that little of the excess wealth is reaching an underbelly of the persistently poor in that region. This means that a net increase in environmental cost is not being offset by a net increase in human wellbeing. But, parallel to this development, the capacity of disaster prone regions such as South Asia to better defend against famine and disease through economic development in the last three decades cannot be disputed.

Dependency and vulnerability

The rationale that economic development reduces disasters by addressing underdevelopment has been a powerful one. With the exception of the disruption caused by the world wars, the transition from pestilence and famine in Europe to relative prosperity during much of the twentieth century and to the present would seem to bear that out. Logically, as rich people get richer, some wealth may pass on to those less fortunate. However, one of the main critiques of the economic growth view of disaster and development is that it assumes a 'trickle down' of wealth to poorer people. A wide body of development scholars (Frank 1966, 1975; Brookfield 1975; Amin 1976; Todaro 1981; Wallerstein 1979; Brewer 1990; Biel 2000) have confronted this assumption with an alternative one which was particularly strong in development studies during the 1970s. This is the underdevelopment school of thought, and, to be more specific, that of the 'development of underdevelopment' (Frank 1966).

The dependency view of development, as it is also often called is based on the understanding that the presence of rich countries is explained by the presence of poor ones. These ideas are often explained through Marxist or anti-colonial writings and are examined in some depth by Brewer (1990). The argument leading to the examination of uneven development and backwardness in Third World countries in the context of world capitalist systems was represented by the work of André Gunder Frank (1976). First, these ideas were largely based on evidence that there had been subordination of economies to those of advanced countries, so that only primary goods were produced for the industrial West. Second, the colonised countries had to maintain an external orientation to meet their dependency on overseas markets for technology, capital sourcing and for production outlets. Third, dependency and the distorted global economy were seen to be in essence further corrupted by the power of elites among whom no sense of equality and fair play could be expected (Hoogvelt 2001).

Some detailed case studies can be found, but one, that by Walter Rodney (1972), famously explained this relationship for the case of Africa, before he was killed by an assassin in Guyana in 1980 (Rodney 1967). Rodney explains how internal development was halted by both unfair terms of trade and the effects on societal evolution caused by this disruption.

Dependency theorists demonstrate how the wealth of the rich part of the world has been built up on the basis of an overall transfer of greater value from poor to rich nations. The colonial history of the world makes it difficult to refute this argument, albeit there have also been other factors to consider. A good critique of contrasting development theories addressing this position from ideas defining developing thinking during the second half of the twentieth century is provided by the edited volume by Ayres (1995), and in a particularly accessible format by Willis (2005) as part of the same series as this book.

Dependency is synonymous with vulnerability, though the latter is not intended here to be used as the substitute for poverty. In disaster management terms we note that a lack of control over personal decision making means that localised strengths in avoiding underdevelopment and major crises are undermined. With the end of European colonial dominance around the world, a post-colonial era was intended to put to right some of the issues of inequality and exclusion that made people dependent. However, the liberation struggles leading to independence rarely ever culminated in peace post-independence. Thus, South Asia in 1948 went through the process of partition between India and Pakistan, with the latter divided into an eastern part, East Bengal, which became Bangladesh in 1972, and a western part, in the Punjab and beyond. The unrest between Hindu and Muslim during this period was associated with what still remains the largest mass movement of people to have ever occurred, estimated as causing some form of displacement to anything up to 40 million people in all. East Pakistan continued to be ruled by external forces, this time from West Pakistan, which in later years tried to impose its language on the Eastern part. A new struggle for independence, what Bangladeshis refer to as the 'second war of Independence', only concluded after some 3 million East Bengalis had died in combat or by war induced famine. Independence and the creation of the People's Republic of Bangladesh finally came about in 1972.

Examples of post-independence struggles in Africa have been commonplace, with almost every country experiencing severe disruption at some point during this period. This was compounded by the Cold War, which meant the superpowers variously funding military power-brokering throughout the continent. Some countries, through their adherence to various versions of African Socialism, found themselves supported by the Soviet Union, Eastern Europe, China and Cuba, whilst others were backed by ex-colonial powers or more specifically the United States. The move to full independence with majority rule for countries such as Zimbabwe, Namibia and South Africa lasted right up until 1980, 1990 and 1994,

Plate 2.3 The National Memorial of Bangladesh. Several million victims of conflict and famine in the 1970s are buried at this site
Source: Author.

respectively. The majority of Portuguese colonies only gained independence in 1975, following the Revolution of Carnations and removal of autocratic powers in Portugal in 1974. As with other newly independent countries that selected a more Marxist approach, a key aim of the new government was to set about nationalising the country's assets and encouraging production. This was with the aim of reducing an entrenched inequality that had persisted under the colonial power, creating a sense of national identity and, it was hoped, bringing more self-sufficient and sustainable food security. Figure 2.2 is a poster from this period that reflects much of the political culture of that time. However, tensions in the global alignment of power between capitalist and communist systems of government, together with many unresolved internal issues relating to sub-regions, tribe or wealth distribution within countries, meant that in many post-independent countries development was curtailed by conflict. Turmoil often led to economic collapse and the need for foreign assistance.

Under the intervention of economic Structural Adjustment Programmes (SAPs), in the ten years from 1981, US$28.5 billion flowed to 64 countries through 187 separate lending operations (Reed 1993). These loans were to fund the restructuring of economies and accompanying systems of governance. The main drive of this neo-liberal agenda was to free up markets for investment from outside, to roll back the protective power of the state over nationalised industries and to make countries become part of the more global capitalist system. To achieve this, most developing countries required large loans. The linking of development to external control through loans from the IMF via the World Bank is often

Plate 2.4 The remains of a Soviet tank in Africa
Source: Author.

Figure 2.2 Poster marking the Third Congress of the ruling FRELIMO party in Mozambique, 1977. In English it reads 'Win the Battle of Production' and then 'Long live the 3rd congress. Long live FRELIMO, the party of the Mozambique people.'

considered as a form of neo-colonialism by the Western powers. Also, by the time the Soviet Union collapsed in 1991, any internationally powerful resistance to a system of global capitalism had been removed.

The problem with Structural Adjustment Programmes in many countries was that there ended up being little evidence that they worked. Furthermore, studies showed that Structural Adjustment Programmes did little to benefit the environment or promote sustainable development (Reed 1993). Closure of state sector industry brought about an increase in livelihood insecurity for many people, and when external private enterprise took over the benefits were often not felt locally. The process and impact of this transition in Mozambique are detailed by Hanlon (1991, 1996) and by Abrahamsson and Nilsson (1995). The wider implications of dependency that debt through aid creates were highlighted early in that process (Hayter 1981; George 1992). The point here in terms of disaster and development is that economic restructuring, foreign investment and debt repayment often led to an increase in vulnerability to disaster.

Once a country becomes dependent on economic development assistance, its endogenous or internal methods of economic survival may be compromised. Rapid change from state to privately owned enterprise can be highly disruptive when it happens too fast, as witnessed in much of the formerly socialist nations. These underlying economic changes are viewed by dependency theorists as the way capitalism has integrated the South effectively into a Western dominated system of exploitation (Biel 2000). A continuum of related issues can prevail between the system of international loans, debt, unfair terms of trade and ultimately recourse to emergency assistance. Middleton and O'Keefe (2001) flag the link to emergency assistance in this aspect of disaster and development, specifically in terms of the politics of humanitarian assistance. In evaluating the process of humanitarian assistance in recent years we find that transfers of aid are political commodities that may help or hinder their intended beneficiaries. They are commodities dependent on the governance context within which aid is received. A noteworthy comment from O'Keefe at the programme sessions at Northumbria University, UK, has been to replace the adage 'a starving child knows no politics' with 'a starving child may only know politics'. This is a reference to the appalling use by those in power of humanitarian assistance to score political points, something that is often understood by local recipients, but not necessarily the donors of aid. It is also a process that has led to the complete neglect of a crisis in some instances and to unwanted assistance in others.

Development aid and humanitarian assistance are commodities that donor nations choose to supply or not dependent on the performances of receiving country. This was reinforced by the UK Prime Minister John Major in the early 1990s by means of the policy of 'conditionality' in foreign assistance. It was a concept also enforced by the IMF, World Bank and other donor organisations or governments. Governments not pursuing adequate market reforms, or where there was evidence of bad governance, could be excluded from receiving foreign aid or humanitarian assistance. Consequently, following Hurricane Mitch in Central America in October and November 1998, a category five hurricane and one of the most destructive ever in the Western Hemisphere, there was some initial hesitation in the flow of foreign assistance due to what was deemed to be the unsuitable governance of the Honduran government. Development aid, and to some extent humanitarian assistance, can therefore become the last chapter of a longstanding political economy of unequal world development.

There is evidence in recent years that some recipient countries are thinking anew about what it is they want to receive and what they do not. For example, following the Indian Ocean Tsunami of December 2004, India decided it did not want to receive foreign humanitarian assistance and that it preferred to manage the response itself. Interestingly, this may have resulted in an increased amount of support to Sri Lanka. It is not clear that the resultant increase in international assistance to Sri Lanka has resulted in more effective aid to relief and recovery than in India. In other instances, humanitarian assistance is wanted but the benevolence of the international community, combined with poor management of donations, may lead to the wrong items being sent. Large quantities of donated items for the great flood of Mozambique in 2000 were sorted into warehouses and could never be used, including milk products, expired medicines and other inappropriate items (Christie and Hanlon 2001). This event was accompanied by an enormous media campaign, perhaps due to its location in the Southern African peninsular during an ongoing period of significant political change in Mozambique and South Africa. Mozambique had recently joined the Commonwealth, the first non-English speaking country to do so, and South Africa had made significant steps towards rebuilding internal integrity and external confidence post-apartheid, through the Truth and Reconciliation Commission. Meanwhile, devastating famines in subsequent years in Niger, Mauritania and parts of neighbouring countries remained underreported and underassisted.

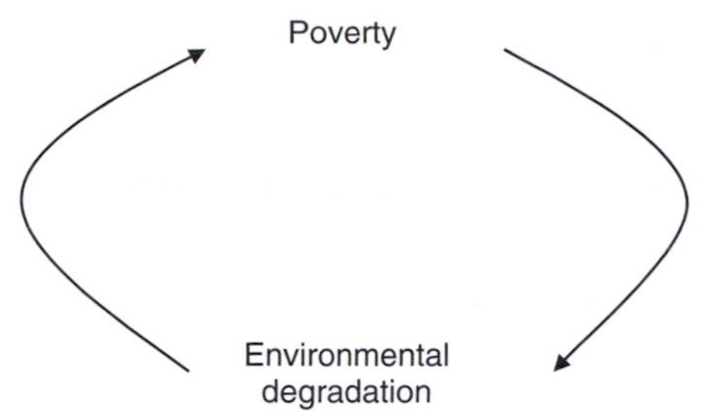

Figure 2.3 Generalised hypothesis on the link between poverty and environmental degradation

Poverty and environmental degradation

The analysis at the start of this chapter based on environmental, social and economic versions of sustainability also provides a clue to deconstructing different aspects of poverty. In this section the term 'environment' includes each of the physical environmental, social and economic components. The links between poverty and environmental disaster can be represented as a downward spiral in which increasing poverty causes environmental degradation, which in turn increases poverty (Figure 2.3).

The manner in which poverty may be both a cause and a consequence of environmental degradation helps unravel this association. The poverty and environmental degradation cycle demands that we consider what is meant by poverty. Three main components are usually identified in understanding poverty. These are income poverty, basic needs poverty and capabilities poverty. The most simplistic form has been defined in terms of income, where a gross national product (GNP) per capita places countries into one of four economic brackets, as used by the World Bank and UN. Absolute poverty is defined against a financial measure, such as surviving on less than one US$1 per day. Relative poverty may also be a financial measure, but one where the figure used is relative to cost of living for a region or local area. Basic needs poverty is more about education levels and availability of health care, as in the Human Development Indexes of the UNDP annual reports since 1997. Capability poverty concerns abilities, access and rights. A more detailed description of a Capability Poverty Index was provided in Chapter 1 (see pp. 20–21). The concept of capability as a core dimension of poverty is also identified by the Development Assistance Committee (DAC) of the Organisation for Economic Co-operation and Development (OECD–DAC 2001) as outlined in Box 2.1.

Box 2.1

Defining poverty through the core dimension of capability

Economic capabilities

These are the capacity and right to earn an income, consume and have assets. They are key to food security, material wellbeing and social status, and may include those assets associated with decent employment, land, implements and animals, forests and fishing waters, and credit.

Human capabilities

These are based on health, education, nutrition, clean water and shelter. Ill health and lack of education are barriers to productive livelihoods and therefore to poverty reduction.

Political capabilities

These include human rights, representation and influence over public policies and priorities. Powerlessness and lack of a voice aggravate other dimensions of poverty in that policy is not reformed and the access issue is not resolved.

Socio-cultural capabilities

These concern the ability to participate as a valued member of a community. They include social status, dignity and the ability to be included.

Protective capabilities

These enable people to withstand economic and external shocks, to be resilient when confronted with external stresses. Insecurity and vulnerability work together as focal aspects of poverty.

Source: Adapted from OECD–DAC (2001).

An evaluative commentary on the DAC guidelines is provided by Middleton and O'Keefe (2001). Qualitative components of poverty have to be analysed using a participatory approach and are sometimes well covered in more localised development practice. The development of this approach as a core perspective and methodology has been progressed and exemplified by the work of Chambers (1983, 1997).

Figure 2.4 suggests that the variable nature of poverty impacts on the quality of the context within which people's wellbeing is determined. Whilst poor people frequently live and work in environmentally poor locations, processes of change involving people who are poor and their environmental context are complex, so that we must again consider

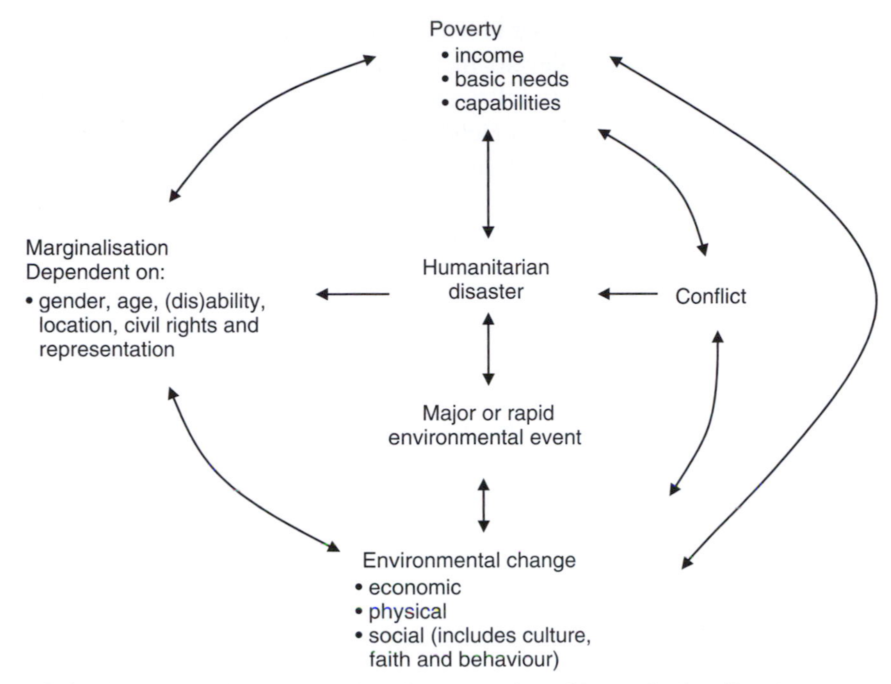

Figure 2.4 An integrated poverty and environment view of humanitarian disasters

physical, economic and social aspects of the environment. Thus, people who lack income, basic technology and access to basic rights, such as, for example, secure land tenure, stay poor. They may exploit basic natural resources unsustainably in order to survive, experience an added social pressure that encourages insecurity, and contribute less to the economy. People in this situation are forced into a cycle, of compromising their investments in health, education and general wellbeing. This impoverished environment leads to the return part of the cycle, which is an increase in poverty, but where the impacts are experienced differently by more or less vulnerable groups. This means that the cycle not only perpetuates itself, but in so doing progressively increases the gap between the rich and the poor, or between the poor and the poorest of the poor.

The monitoring of poverty has frequently shown that female headed households are disproportionately represented amongst the poorest groups. The gendered dimensions of poverty have been well documented by almost all of the international organisations and NGOs in recent years. Data associated with development projects are now frequently aggregated for gender as a result. The OECD–DAC (2001) developed its entire breakdown of the interactive dimensions of poverty and wellbeing around the link between gender and environment. However, it is important to remember that the manner in which this relationship has been understood by different cultures, development scholars and personnel has been frequently reanalysed (a full account is provided by Momsen (2003) as a part of this book series). A Gender-related Development Index (GDI) and Gender Empowerment Measure (GEM) are available from UNDP (1995). It ranks Norway as top and countries such as Mali, Nigeria and Kenya around the bottom in terms of this index.

Apart from gender, people become marginalised on account of age, disability, religion, access to civil rights and representation, location or disease status (such as being HIV positive). This distribution may also be reflected across a spectrum of rural and urban development, as is emphasised on further in the subsection of this chapter that follows. The process of increasing environmental degradation, including badly degraded rural or urban environments, economic fragility and social decline, is a progression of vulnerability that increases disaster risk. The downward spiral of poverty and environmental degradation continues until disaster strikes, and in this version of events we would often expect an increase in poverty to result. There are alternative scenarios that can be recognised, particularly those of Figure 2.5. But if we for the moment consider the process of famines in Africa and formerly in South Asia, and of displacement and impoverishment in and around most of the developing world's major urban areas, of unnecessary disease epidemics and of infrastructural disasters, there is a worsening of environmental context with poverty. The slow onset process of poverty and humanitarian disaster is speeded up more critically by the onset of conflict and rapid onset disaster events as depicted in the diagram.

Reversing the cycle

Having looked at a version of events that indicates the progression of poverty, increased environmental degradation and disasters, it is important to demonstrate that the process can be set in motion in the opposite direction. In the scenario presented in Figure 2.5, wealth and wellbeing lead to a process of improved inclusion and security for people, such that the environmental context improves and poverty is reduced. Conflict mitigation is likely to be more effective in these circumstances and therefore destabilisation in that way less likely. The process of improvement in inclusion, cohesion, rights, particularly for the mass of people who are poor, in turn also leads to reduced vulnerability to disaster. Absence of disaster, as we shall see further in Chapter 3, also improves livelihoods, environmental quality and the economy, and these in turn reduce the vulnerability to disaster.

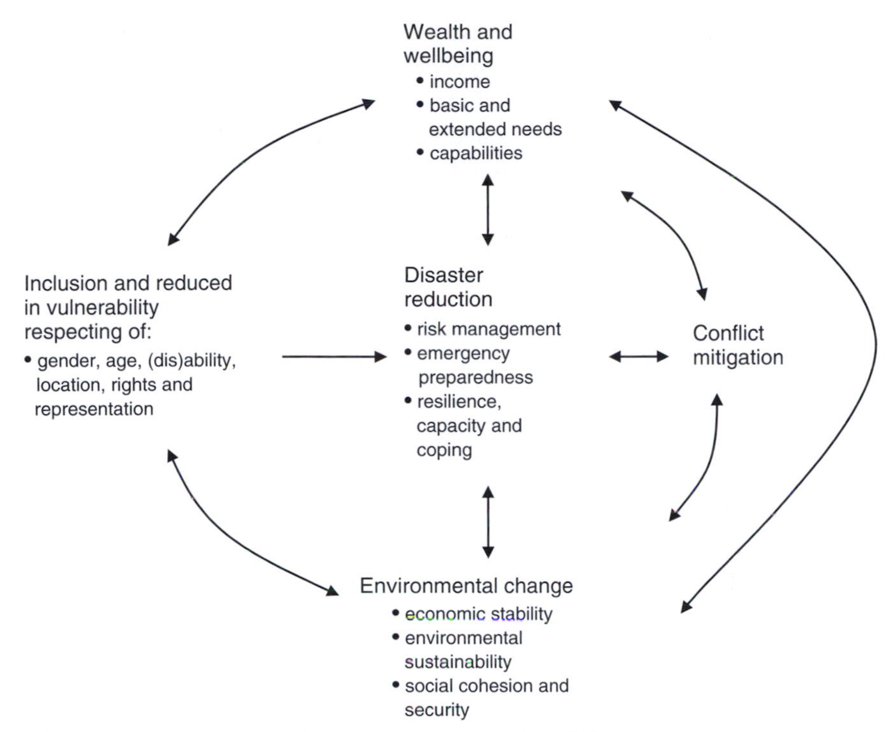

Figure 2.5 Idealised reversal of poverty, environment and disasters processes

Poverty alleviation

An integrated view of processes of poverty, not to mention its implications for disaster risk, has led to integrated poverty alleviation strategies. These are apparent at all levels of intervention. We can see them at the macro-policy making level. The MDGs represent an integrated strategy towards reducing poverty. More specifically, we have the integrative approach of country-level PRSPs, which represent a range of government sectors and detail how they contribute to the poverty reduction agenda. These must also reflect priorities of the World Bank and other donor views on poverty reduction, but also those which are considered of concern at national level. In recent years, there has been a remarkable convergence of the rhetoric and apparent intention of development policies at international and national levels. This suggests that what we are now seeing is a globalisation of thinking regarding international development, and a similar process might be expected for disaster reduction policy.

Achieving common goals across the nations in this way is considered a significant success, but some caution is also felt in that it is local solutions that are often needed for addressing global problems. No two local areas are the same, such that they are not reachable through generalised global-level policy speak. Potential dilemmas in poverty alleviation at the macro political level have been for decades about whether a nation should pursue growth-mediated security or support-led security. The growth-mediated version concerns the greater role of the private sector, trade liberalisation and rolling back the influence of the state, as in the SAPs we discussed earlier (see pp. 67–69). The support-led version concerns state investment in services that reduce poverty and, where necessary, subsidies to protect assets on which people rely. Despite so much progress in development over the last few decades, not least the rapidly emergent economies in East Asia, in practice macro-scale versions of poverty reduction have failed to eradicate it.

The shortcoming of not including people in the design of their own poverty reduction strategy has been considered one of the main constraints in development policy of the past. It has led to the emergence around the 1990s of what we consider a new paradigm, that of participation through grassroots, bottom-up development strategies as advocated by Chambers (1997). Whilst it seems instinctively correct that people participate in their own development and disaster reduction, the approach has nonetheless drawn concern from some quarters. This is because participating does not necessarily mean being in control. What is really required is participation that comes from empowerment. The difference in practical terms is whether a local community is just a part of the process of deciding development and disaster reduction strategies or whether it decides on and owns that process. Participation could also be seen to present an undesirable opt-out clause for governments, who should still maintain responsibility for many roles. Some of these issues are discussed further by Francis (2001), in terms of the World Bank view, and by Cooke and Kothari (2001), on issues regarding its dominant influence in recent development work. The techniques of participation in disaster and development work are returned to in more detail in Chapter 5. Many of the dilemmas of development work in terms of participation, empowerment and the nature of intervention in development come down to whether there is sound communication with and facilitation of existing local capacity. If progressed in response to a well-understood need, where there is existing motivation to enhance a strategy, then success in poverty reduction is more likely.

Having a sound policy context within which local solutions might develop is an advantage. For example, in 1997 the United Kingdom Department for International Development launched its own integrated specific objectives for poverty reduction, as follows:

- policies and actions which promote sustainable livelihoods;

- better education, health and opportunities for poor people;
- protection and better management of the natural and physical environment.

(DFID 1997)

The first of these is a broad topic that has reflected and encouraged the development of a number of interesting strands of development thinking. It is an approach to facilitating understanding of people centred poverty assessment and reduction, and of sustainable development:

> A livelihood comprises the capabilities, assets and activities required for a means of living. A livelihood is sustainable when it can cope with and recover from stresses and shocks and maintain or enhance its capabilities and assets both now and in the future, while not undermining the natural resource base.
>
> (DFID 2000: 1)

This definition is particularly revealing about the emergence of disaster and development thinking, in that it refers to livelihood sustainability in terms of coping and recovery from stresses and shocks, terminology which is increasingly commonplace in disaster reduction. Various livelihoods frameworks have been produced in recent years, one of the most frequently referred to being that of DFID (2000), elements of which are referred to further in the following chapters of this book. These are the frameworks and approaches that define a programme of activity. A further version of a livelihoods framework is the one used by the NGO CARE International (www.careinternational.org.uk). They are readily available from the websites provided in this chapter (see pp. 121–122). A local reinterpretation is used from a poverty reduction project engaged with by the author (Figure 2.6). This was supported by DFID and the Government of Mozambique in Zambezia between 1998 and 2003, and was broadly situated within the DFID livelihoods programme that had just been established. Figure 2.6 was the project's own interpretation of how various aspects of the livelihoods process could be best represented at the household level for the target population. In this representation, an idealised process of household wellbeing is locked into a process where capabilities, assets and access to production are made use of through development of household-level strategies. Planned strategies become activities that produce the positive outcomes indicated. These in turn improve household capabilities, assets and access. However, this localised livelihood process occurs in a context where vulnerability inducing factors such as disaster shocks, unfavourable development trends and seasonal adjustment put pressure on the household and institutions that influence households. Whilst the household is able to

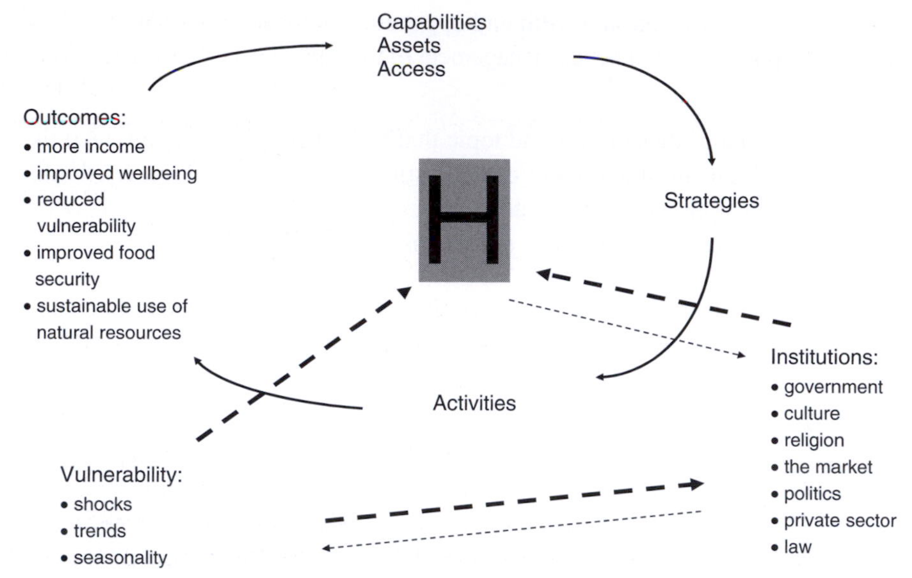

Figure 2.6 Household sustainable development for poverty reduction, where H = household

mitigate some of the impact, the process of livelihood security is disrupted. This is explored in more detail in Chapter 3, where consideration is given to how disasters impact on a wide range of human assets. Referring to the DFID livelihoods framework, these include natural, physical, social, human and financial capital. A description of the development of the ZADP integrated poverty reduction project based on at least some of these aspects is presented as Box 2.2.

Urbanisation and globalisation

The influence of development issues on disasters is apparent through the process of urbanisation and globalisation. Environmental, social and economic changes may be intensified and disaster risks may change more rapidly in those contexts, particularly in developing regions. Changes in the economy that affect the opportunities for sustainable livelihoods in rural areas lead people to migrate to urban areas, where the birth rate may already be accentuated. Where there are more people and a rapidly changing environment there is more exposure to hazards and a greater range of immediate risks. Whilst the economic development of the newly emergent economies of East Asia such as Thailand is associated with a decline in some infectious diseases, poisonings and accidents

Box 2.2

Zambezia Agricultural Development Project (ZADP), Mozambique

This project represented an attempt to address the potential of famine in what were in the 1990s three of the economically poorest districts in one of the economically poorest provinces of one of the world's economically poorest countries. The overall context of poverty in this area was determined by thirty years of conflict that ended in 1992 and which decimated the rural areas through mass population displacement, a complex development history before and after independence from Portugal in 1975, and occasional but severe droughts. However, during some periods of less conflict, pre- and post-independence, this area had also managed to be relatively productive in terms of agriculture. The project sought to find ways of reducing the vulnerability to famine and consequent dependence on emergency assistance through addressing key influences on food and livelihood security. Participatory appraisals were carried out in some of the worst affected areas to facilitate a community based needs assessment. Following further analysis, the following problem tree was established which then guided the planned activities of the project:

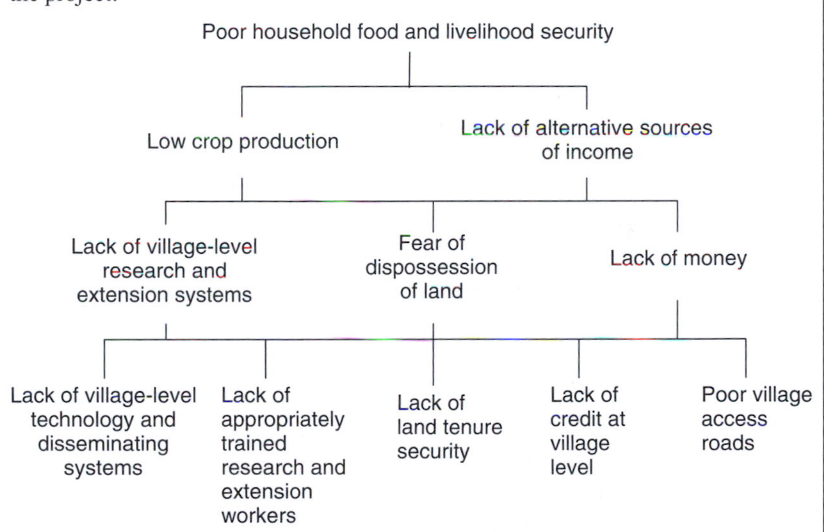

The main objectives of this integrated approach to poverty and disaster risk reduction therefore translated in practice to:

- strengthening farmer centred research and sustainable farming systems for improved agricultural output;
- developing sustainable systems of credit for the poor;
- developing systems that enable smallholder producers to secure land tenure.

It also fitted the policy and programme focus of DFID, with an approach to food and livelihood security in line with the strategy for poverty reduction that aimed to assist poor people establish sustainable livelihoods (DfID 1997: 19). Specific features of the project that it was considered possible to address with the 7.5 million funding were to:

- increase farm production and income sources for the rural poor in three target districts;
- give priority to the poorest households, a significantly higher proportion of which were found by both participatory appraisal and the project baseline survey to be female headed;
- target households in less accessible areas, which are often poorer;
- be a multifaceted approach to improving household food security;
- facilitate impacts on food and livelihood security, such that neighbouring areas would adopt the successful aspects of its work in these target areas.

There were many successes of this project, but also many challenges surrounding what it attempted to achieve. Some of these included the difficulties of:

- defining poverty reduction in terms of its impacts across the assets, access and capabilities confluence;
- deciding whether in this context it was a greater priority to target food or livelihood insecurity;
- being a grassroots oriented project but having to operate through top-down structures in its cooperation with government;
- targeting the poorest of the poor, whilst it is more likely to be the economically active people that bring growth and development;
- balancing donor and implementing agency policy (in this case World Vision) with more complex experiences from the field level.

Wider ethical issues included an ongoing debate as to whether it was possible to effectively intervene in such a delicate and localised development situation, and the question of how activities would continue once the project left. As part of the project the rights and representation issues of the intended beneficiary communities were engaged with, particularly in the case of land tenure. Yet, despite some success, questions remained as to whether a development programme of this type can really improve rights and representation issues, which are either locally grounded matters or are controlled by government.

Source: ZADP Project Documents, 1998–2003
The author worked as this project's external Monitoring and Evaluation advisor for periods throughout most of the life of the project.

rapidly increased in the 1980s (Philips and Verhasselt 1994) due to urban development. That health is the largest category of hazards is also reflected in the case of urban areas.

When urban development attracts the poor, whose vision of building their livelihood there is not achieved, there is an increase in the number of vulnerable people forced to live in hazardous locations. Shanty towns around the cities of South America, Asia and Africa exemplify this, as the poorest of the poor find themselves with only the option of living on the steepest, most unstable plots, or next to sources of industrial or household pollution. All cities have a proportion of people with no home at all, and in some developing world cities a sizeable part of the population may be physically displaced in one way or the other. Meanwhile, the process of urbanisation where there is a more vibrant macro-economy is linked to an unremitting globalisation that can drive forward development with little concern for its marginalised urban poor.

Bull-Kamanga *et al*. (2003) point out that the definition of a disaster is important to looking at urban disasters in that what is considered exceptional in a developed world city, such as infectious disease deaths, is considered normal in the developing world. They point to the continuum that exists between major disasters, 'small' disasters and everyday hazards, suggesting that very frequent small impacts from everyday hazards are actually the same as low-frequency big impacts from major events. Another point they make is that 'the populations of smaller urban centres are often at greater risk than those in larger urban centres' (Bull-Kamanga *et al*. 2003: 199), as water sanitation, drainage and health care are worse. This is at least the case for many of the African cities, which are generally not as large as those in Asia, South America or Europe. Alexander (2000) sums up some of the urban disaster terrain as follows:

> The world's largest cities are burdened with the threat of hybrid risks composed of mixtures of natural extremes, technological failures and harmful social activities. In addition, urban conflict often complicates the assessment and relief of disasters, and not merely in the cities of poor, developing countries. At the same time, changes in the pattern of urbanization, for example suburban expansion and inner-city decay, result in shifts in the locus of hazard and risk. Thus, impacts and costs are rising steeply amid highly uncertain social, economic and technological contexts. The problems are especially acute for major Third World cities, which often suffer a shortage of the most basic scientific information on hazards.
>
> (Alexander 2000: 95)

Large numbers of urban dwellers live in poverty. However, as highlighted in the case of South Asian cities, 'the intensity and depth of vulnerability varies among them . . . the extremely vulnerable not only live with a perpetual sense of uncertainty and fear but also lack support from their families and the community' (Bhatt 1998: 23). Looking specifically at those in the Indian informal sector belonging to lower Hindu caste groups and the Muslim community, the women were found to be vulnerable for a wide range of reasons. Bhatt points out that 'respondents were able to identify that, even while being vulnerable, some of them were more vulnerable than the rest and some were near-victims or destitutes' (1998: 20)

Although a developmental view of disaster essentially emphasises the marginalisation of the sub-groups as a precursor to disaster, it may be that some urban risks are not so much about who you are, but where you live. For example, a calculation of vulnerability was carried out by the author across the sub-areas of three of Mozambique's urban areas, including the second largest city of Beira. Conditions were so hazardous in terms of poor water supply, sanitation, drainage, general levels of survival and exposure to infectious disease hazards that data suggested that it made little difference in terms of immediate vulnerability whether a household was comprised of war refugees or long term residents of the town (Collins 1998). The people living in these circumstances suggested that there was a lower threshold of wellbeing, below which exposure to hazards was no longer any different for people displaced from the rural areas than it was for the long-term residents.

However, a return study of the city of Beira in 2004 found that there was a tendency for lower 'wealth groups' to be overrepresented amongst those experiencing greater numbers of crises. This was by now during a period of relative economic development compared with the earlier study. An explanation might be that refugees compensated for their circumstances with their additional resourcefulness, or that differentiation in wellbeing between them and long-term residents was erased by the overall upward shift in economic development at that time.

There are texts that highlight the staggeringly precarious status of the world's cities in terms of potential crises, and suggest solutions. A particularly useful account of the vulnerability of cities to disaster has been provided by Pelling (2003a), who provides case examples to explain how resilience in the city is in part a function of its system of governance. Hardoy *et al.* (1992) have focused in some detail on the environmental problems. Urban planners have developed ideas of urban sustainability

Plate 2.5 The recently expanded cityscape of Kathmandu, Nepal. It is in an earthquake zone

Source: Author.

using ecological models, such as that of the Dutch 'Ecopolis' (Tjallingii 1995). This can be summarised as development that is based on the regulation of flows for a responsible approach to use of resources, a coherent spatial planning aspect with open areas for a 'living city', and participatory approach in terms of lifestyles and economic activities (Selman 1996). This is managed, however, by having clear and responsive administrative levels which allow for top-down rationalistic perspectives alongside community based development activities.

Whilst urban areas are centres of commerce and their expansion is a reflection of global economic development and changes in wealth, over-population and intensified risks to life in many cities of the developing world are often linked to poverty in rural areas. Meanwhile, urban environments are increasingly part of a globalisation process, and it is this that may exacerbate the overall human induced physical risks they are associated with. Risks are greater for those less able to adapt to urban areas or who are surplus to employment needs. In most urban areas, particularly in the developing world, an underclass of people are exposed

Plate 2.6 One small corner of urban development at Beijing, China will bring many from the rural areas to live in these apartments
Source: Author.

to more hazardous living conditions that compound their socio-economic vulnerability. Understanding the role of globalisation in increasing disaster risk requires careful analysis. In general terms it represents what Oliver-Smith points to as an increase in ecological, social and economic 'flows of energy, information, material and people, intersecting densely at various points and less densely at others around the globe' (2004: 22). What can be added here is that dense points of interaction that increase the chances of these forces colliding either constructively or destructively are generally urban in nature.

Complex political emergency

So far, this chapter has established that disasters can be caused by multiple aspects of development. The complexity is derived from there being environmental, social and economic origins of disasters, which may be dependent upon actions at institutional, community and individual levels. We have also emphasised that disaster risks may work together, and that they are compounded during certain periods or in concentrated areas. By understanding disaster relative to varying perceptions that people have about it, which may reflect cultural and socio-economic differences, the complexity is increased. However, the term 'complex political emergency' is more frequently associated with conflict and open warfare. Certainly, everything that has been covered so far in this chapter may have a political element, in that policy is derived, contested or subsumed in environmental, social and economic aspects. The complexity of disasters has been recognised by most authors on disasters in one way or the other, including Rosenthal on disaster 'characteristics, conditions and consequences' (1998: 150) and those that 'occur without a clear-cut single natural hazard trigger . . . or identifiable event in the political economy' (Wisner *et al.* 2004: 91). Complexity and politics have also been identified as part of the humanitarian aid dilemma in responding to disasters (Pirotte *et al.* 1999; Middleton and O'Keefe 1998).

Conflicts have not always been considered part and parcel of the disasters community remit, much of which has developed from the 'natural hazards' strand of the subject field. However, those who must respond to a flood, fire, infectious disease emergency or terrorist incident tend to be the same, namely the emergency services. Humanitarian responses sometimes comprise similar aid, human resources and strategies, regardless of whether people have been displaced en masse by war or by a major environmental event. With greater emphasis on a risk reduction approach in recent years the risk reduction for environmental and technological disasters might require some of the same preparations politically as those required for conflict risk reduction. The relationship between more stable and unstable environmental contexts, of security, politics, entitlements and social capital, is addressed by Goodhand *et al.* (2000). The examples that have provided the rationale for viewing disasters as complex political emergencies have often been based on the case of the Horn of Africa region, where the effects droughts are compounded by politics and warfare. An overview is provided by White (2000) as a precursor to a special edition of *Disasters* on this topic in 2000. More recently, it has become evident that terrorism is closely linked to complex emergencies and to how disaster management policies are funded. It is no secret that in the UK the Civil Contingencies Act lost much of its focus on issues such as the Foot and Mouth Disease crisis and flooding once there had been terrorist attacks in the US and then the UK in the first half of the 2000s. The FEMA budget in the USA may have been diverted from preparedness for environmental emergencies, such as Hurricane Katrina, by what appeared to be a more pressing Homeland Security Agenda in the preceding period.

Complex political emergencies compound contemporary disasters. The challenge has increased where development induced disaster is conflict induced and complex on account not just of political economy but also of juxtaposed cultures with uncertain futures for commonality.

Conclusion: disaster management through sustainable development

The nature of the link between development and disaster varies for different notions of what constitutes a disaster, including individual or complexes of environmental, technological, economic, social and political changes. Views have tended to be either more determinist fatalistic, based on apocalyptic no-hope scenarios, or more possibilistic, whereby it is believed that the right type of development avoids disaster. Analysis of sustainable development across its environmental, social and economic dimensions suggests that disasters are conceptually not inevitable. They can be managed through an integrated vulnerability reduction approach. Poverty reduction may not be in its entirety the same as vulnerability reduction, but it nonetheless lies at the heart of reducing a large cohort of disaster risks.

Development indicators of disaster risk are not just comprised of income and basic needs factors, but also of rights, representation and absence of conflict. Pressures of urbanisation and globalisation have created a more risky world for which we are challenged to urgently reassess development for both local and global disaster reduction. Identifying what type of sustainable development, who decides and how to most effectively implement it will remain as complex as development itself, and is a means to disaster management.

Discussion questions

1 Is there really a global over-population crisis, or is the threat of disaster a function of poor management of resources? Relate population and environment debates to different perspectives on sustainability and disaster.
2 Are tragedies of the commons really tragedies of enclosure? What are the implications for disaster risk of either (1) allowing private ownership of environmental resources or (2) declaring all environmental resources part of a global commons that nobody can own?

3 What are the real causes of poverty and development induced disaster risk? What indicators are commonly used to measure poverty. How useful are they in highlighting who is vulnerable to disaster, and the reasons for this?
4 What are the alternatives to seeing people as vulnerable? How do people address their vulnerability to disasters? How do different development policies aim to reduce poverty and consequently the risk of disasters, and how effective are they in achieving that aim?
5 Can sustainable development always reduce the risk of disasters and at what point does disaster management become an issue beyond development alone?

Further reading

Adams, W.M. (2001) *Green Development*, third edition, London: Routledge.

Allen, T. and Thomas, A. (eds) (2000) *Poverty and Development into the Twenty-first Century*, Oxford: Oxford University Press.

Elliot, J.E. (2006) *An Introduction to Sustainable Development*, third edition, London: Routledge.

Hoogvelt, A. (2001) *Globalization and the Postcolonial World: the new political economy of development*, Basingstoke: Palgrave.

IFRC (2004) *World Disasters Report*: *Focus on Community Resilience*, Geneva: IFRC.

OECD–DAC (2001) *Poverty Reduction: the DAC guidelines*, Paris: OECD–DAC.

Willis, K. (2005) *Theories and Practices of Development*, London: Routledge.

Useful websites

www.eldis.org: portal for development related information run by the Institute of Development Studies, University of Sussex.

http://www.undp/povertyhome: United Nations Development Programme (UNDP) site on poverty.

http://www.worldbank.org: World Bank.

http://www.ids.ac.uk: Institute of Development Studies, University of Sussex.

http://www.dfid.uk: Department for International Development (UK).

http://www.livelihoods.org: on sustainable livelihoods.

http://www.odi.org: Overseas Development Institute.

http://www.oecd.org/dac: the Organisation for Economic Cooperation and Development, Development Assistance Committee (OECD–DAC).

3 How do disasters influence development?

Summary

- **The effects of disasters on development are dependent on underlying development trajectories and the intensity of different hazards and vulnerabilities.**
- **Disasters alter social and economic terrains, potentially resulting in both development losses and some gains.**
- **Disaster reduction involves human coping, resilience and security as ingredients of wellbeing and sustainable development.**
- **Periods of crisis generally involve displacement, through either forced migration, destruction of places (topocide) or people's loss of connection to places.**
- **Disaster relief influences the governance of national and local development.**
- **The way major emergencies are represented affects national and international development priorities and assistance.**
- **Proactive disaster prevention strategies reduce disaster losses whilst enhancing longer-term sustainable development.**

Introduction

This chapter considers how disasters influence development. We saw in Chapter 2 explanations of how we can interpret development influences on disaster in terms of hazards and human vulnerabilities. The disaster and development nexus is part of a circular process within which the impact of disasters on development needs to be further analysed. Knowing about the impact of disasters on development can assist in identifying the fuller meaning of disaster mitigation and avoidance. This is because disaster reduction initiatives can be invested in better if they are understood in terms of their wider and longer-term benefits.

Both rapid onset and slow onset disasters have impacts on development. Large and sudden events, such as the Indian Ocean Tsunami of 26 December 2004, have an immediate and widespread devastating impact. The brunt of the impact of the tsunami was rapidly experienced in Indonesia, Thailand, Sri Lanka, India and the Maldives, whilst countries bordering the Indian Ocean as far away as Somalia and Kenya were affected. When major hazards occur more frequently, such as the annual flood cycle of South Asia, adaptation to crises may be longer-term and more effective. We have tended so far to think of the relationship as cyclical because development problems provoke disasters and disasters slow up development, which then causes a chain reaction that can perpetuate this cycle. However, the relationship does not in practice progress in so systematic a way as rates of change and the extent of areas associated with disaster onset and recovery are very variable. Different types of disaster, such as, for example, drought, flood, earthquake and epidemics, have variable influences on development.

Development changes may include specific defences against major disasters. For example, the annual flood and cyclone hazards in Bangladesh have led to the establishment of a flood defence system and cyclone shelters as part of government development policy there. Donors to the 1991 cyclone allowed the humanitarian agencies, the Bangladesh Red Crescent Society and the Government to create risk reduction infrastructure lacking since the major catastrophe in the 1970s that took over 500,000 lives. The cyclical disaster and development analogy may be less useful in depicting crisis as part of an overall development situation with combinations of immediate and ongoing underlying issues. Risks are continuously changing, whilst even during a period of emergency such as a flood, tsunami aftermath or security breach, development policy undergoes review. A disaster will alter national development budgets, internal political alignments and international relations in both the very short and the longer-term. Thus post-disaster reconstruction becomes a development process to the extent it is consistent with more regular development policy.

A typology of disasters and their impact on development

Disasters by definition cause loss of life, livelihood and wellbeing, consequently stalling development or setting it on a backwards trajectory for months, years or decades. However, a typology of disasters and development linkages demands that we take a closer look at the nature of a disaster to interpret its varied impacts on development. As introduced in Chapter 1, and highlighted through the group contributing to the Quarantelli (1998) text, existing data on disasters reveal a situation of contrasting nomenclatures. The explanation is in part that disaster data are recorded for varying purposes: disaster prediction and prevention, monitoring and evaluation, mitigation and impact assessment, insurance and so forth. The recording of disasters alone is not consistently carried out, and therefore comparison of disaster and development linkages around the world is not a consistent exercise. Guha-Sapir and Below (2006) found there are wide differences in disaster data between the four countries they selected: Vietnam, India, Honduras and Mozambique. Their comparison was made using NatCat data, maintained by Munich Reinsurance Company (Munich); Sigma, maintained by Swiss Reinsurance Company (Zürich); and EM-DAT, maintained by the Centre for Research on the Epidemiology of Disasters (CRED, Université Catholique de Louvain, Brussels).

Having a reasonable understanding of what the data are trying to show is important in terms of planning and costing the likely impact of forthcoming disasters. Otherwise it might be argued that the different categorisations of data in relation to their purpose might matter less so long as broadly consistent methodologies for acquiring data and common definitions are adhered to. There is currently no evidence available to suggest this has been the case. Nonetheless, a growing concern with disaster reporting may be gradually bringing estimates closer to actual numbers so that more accurate comparisons of disaster with development are feasible. It is not the intention of this book to systematically present the standard disaster data, as these can be easily accessed by the reader from EM-DAT, or via the ICRC outlets indicted at chapter ends. Rather, we must pause here to look at what categorisation and types of data might be useful in assessing an impact on development.

The basic classifications of disasters based on Hewitt and Burton (1971) list atmospheric, hydrologic, geologic, biologic and technologic hazardous environmental elements. If we are using IFRC figures based on EM-DAT we normally find categories with ‘a natural and technological trigger only', not including 'wars, conflict-related famines, diseases or epidemics' (IFRC 2004: p.173). However, even if using IFRC hazard types from 2004 alone (rather than risks based more exclusively on vulnerability), Taylor (1989) and then Tobin and Montz (1997:10) would extend these to 'natural, industrial and human ones'. They can be subdivided across those associated with 'Earth, Air, Fire, Water and People'. The hazards here can be typically based on a diverse range of earthquakes, storms, urban smog, design flaws, transport accidents, disease epidemics and terrorism.

K. Smith (2001) indicates that disaster events can be arranged in terms of their 'naturalness' or as human induced, in terms of involuntary or voluntary, or intense and more diffuse. However disasters are defined, the impact on development of such a diverse range of phenomena will clearly be very variable, particularly as we know disasters to often comprise multiple components. Categorising disasters more in terms of their longer-term development impact may be more useful than simply describing them in terms of an environmental event and numbers of casualties. It should be appreciated that the loss of just one individual or 10,000 may be equally grave a disaster experience to a bereaved individual. Assessing a disaster in terms of development is a more relevant way of understanding its immediate and extended costs.

In terms of natural hazards and the problems they cause, the following working definition supplied by Smith is useful in making it clear that disasters are best defined by their impact. Environmental hazards are

> Extreme geophysical events, biological processes and major technological accidents, characterised by concentrated releases of energy or materials, which pose a largely unexpected threat to human life and can cause significant damage to goods and the environment.
>
> (K. Smith 2001: 17)

Putting a focus on varying human vulnerabilities aside for the moment, and with reference to Hewitt and Burton (1971), disaster impacts are seen as dependent on their areal extent, intensity, duration, rate of onset and predictability. Benson and Clay (2004), writing for a World Bank study, distinguish the differences of impact between hydro-meteorological and geologic hazards. Based on the cases of Dominica, Bangladesh and Malawi, they conclude that the 'recurrent nature of hydro-meteorological hazards encourages adaptation in economic and social activity. Geophysical hazards, by contrast, are mostly low probability, seemingly random events, and their risks are often almost wholly discounted'

(Benson and Clay 2004: 61). But, the financial costs of these impacts are rising with economic development.

Although some costs can be identified, the variable impacts on more or less vulnerable people and the effects over time make it impossible to fully cost the impact of most disasters. Whilst rains were still falling on the floods of central and southern England during July 2007, estimates of more than 3 billion of damage were being made. However, the real economic costs are unknown, let alone other social and psychological costs. House prices in flood prone areas may decline over a long time-frame following this type of event, whilst insurance premiums are likely to rise. On the other hand, promised investments by government in better flood defences might counteract some impacts, though they in turn may have an unpredictable development result. Beyond the financial and tangible costs lie many intangible effects. Direct impacts of environmental disasters are when a building collapses, a crop is lost and people are killed or injured. Indirect impacts are the wider implications of the loss of a livelihood, redirection of finance, loss of confidence in a market, psycho-social impacts, and potential loss to heritage and way of life. These losses unfold to their full extent over the longer time-frame and are often only experienced well after the immediate costs of disaster have been broadly estimated.

A further example is presented by the losses experienced through the Indian Ocean Tsunami. These were catastrophic in terms of impacts on human life and livelihood, the UN indicating that there were 229,866 people dead or missing. In addition to it being such a catastrophic rapid onset disaster, the longer-term impact has been a fragmentation of communities. One explanation for this in the case of Sri Lanka has been the influence of foreign aid and relief in the coastal areas, which could not be controlled even-handedly. Though this is less obvious in the short term, and unintentional, an infusion of assistance can increase disparities between rich and poor and divisions between different ethnicities. K. R. Smith (2001) and others, however, have pointed out that disasters bring some tangible and intangible gains. These would include any hidden benefits of reconstruction grants, or the gains in environmental terms through improved fertility of soils, or through subsequent opportunities for new housing, infrastructures and tourism in newly regenerated areas.

> A disaster with all its negative consequences offers a good opportunity to formulate forward-looking policy concepts pertaining to social development and equity, economic growth, environmental quality and justice, i.e. sustainability.
>
> (UNISDR 2004: 21)

Plate 3.1 An Internally Displaced Persons (IDP) camp in Eastern Sri Lanka
Source: Janaka Jayawickrama, Northumbria University.

Loss, opportunity and exploitation through disaster

Multiple hazards and vulnerabilities contribute to an overall loss in development terms. In line with the view of development used in this book, losses will be broadly in environmental, social and economic dimensions. In terms of environmental sustainability, loss is where ecological systems are disrupted or decimated, biodiversity diminished and the means to regeneration curtailed. Losses may include those caused by mass pollution, such as through a chemical leak, oil tanker spill, or more subtle environmental changes such as the increase of arsenic in a water supply. Slow or longer-term environmental losses are currently being associated with climate change. Being intangible in the short term, the losses start to be realised as floods, cyclones and droughts are increasingly aligned with what are thought to be human contributions to climate change. However, environmental losses through climate change are far from clearly categorised. Overall, the physical environmental changes that we may predict with climate change are no greater than the world had experienced previously, through ice ages, blackouts from super-volcanoes, colliding meteorites and mass extinctions. The difference, therefore, is not in terms of the impact on the environmental system, which is likely to recover, but on human survival through dependence on regularity and reliability in environmental systems. Looking at it this way, the extent to which people can continue to adapt to environmental changes through modifications to food, shelter, fuel and processing of waste is the extent to which there is longevity in human survival. It is no surprise therefore that the Intergovernmental Panel on Climate Change (IPCC) has since 2002 stressed human adaptation to climate change as key to sustainable survival strategies.

The contemporary issue is tied to circumstances of there being so many people dependent on the survival of such a susceptible range of environmental resources. Small changes in the environment, in terms of its feedback on ecological systems, can represent very large changes in people's options for harvesting resources. Just one degree increase in temperature threatens droughts in Africa, sea level rise and greater flooding in South Asia. Environmental losses of disaster can therefore also be thought of as changes in environmental systems for which people are currently ill adapted. Also, impacts through increased climatic hazards or mal-adaptation in one part of the world can no longer be isolated from other parts of the world. We live in an age of global risk, as environmental inputs and outputs in one region affect another, whilst flows of finance, aid, trade, people and health hazards follow vulnerability around all over the world.

Social aspects of loss with disaster include loss of immediate family members, relatives and friends, of people that make up the coherence of a community, its leaders, those with memories of the past, institutional memories, of skills and diverse abilities. Some disasters selectively remove particular categories of people resulting in a skewed social impact. For example, cyclone disasters around the Bay of Bengal have been found to result in more female deaths than male, as women stay in the homes more and these are more vulnerable locations. Industrial disasters, such as the Piper Alpha oil disaster, impact on the families of niche category workers. Many industrial disasters, such as the Bhopal chemical plant leak in India in 1984 or Chernobyl nuclear facility in Ukraine in 1986, impact on all residents in the vicinity. In such instances, loss becomes area based. However, in arguably the world's biggest disaster zone, that of infectious disease epidemics, particular age groups are selectively more at risk. Thus, under-5 infant mortality is used as a development indicator by the UN and other international organisations, the greatest contributors being malnutrition, diarrhoea and malaria. The losses from HIV/AIDS in Africa, which account for more than 80 per cent of cases, mean that survivors are disproportionately the very young and the old. The development impact is brought about by loss of the economically active part of society. This has also brought into play new social alignments, such as child-headed households, grandparents acting as the parents of their grandchildren and widespread adoption (Barnett and Whiteside 2002).

Economic losses are more often calculated at a macro level, whilst costs are experienced at the more local or individual level. Economic estimates of loss are exemplified in those parts of the world where insurance calculations are made. Tobin and Montz (1997) cover examples particularly for the United States, where there are specific insurance initiatives for disasters, such as the Flood Insurance Scheme. They also outline economic impacts of disasters more widely and indicate that there is a play-off between people's reliance on federal disaster relief, insurance and other community response options. The costs in relief, however, seem to be rising in the economically developed world. The greatest relief allocation in the United States between 1984 and 1994 was 2.9 billion for the Northridge Earthquake, California (Tobin and Montz 1997: 249). The full economic cost of Hurricane Katrina in 2005 has not yet emerged, but is likely to be much greater. Even in cases where loss of life has been relatively low, such as the UK floods of 2007, the economic cost of flood damage was estimated at more than 3 billion (6 billion). Benson and Clay note that 'natural disasters can create significant budgetary pressures, with potential narrow fiscal impacts in the short term and wider long-term implications for development' (2004: 62). The primary way in which this occurs is through reallocation. They indicate that the brunt of reallocations appears to fall on capital expenditure and the social sector, that there is also considerable in-kind reallocation of human and physical resources, but that this diversion of development investment is in most instances poorly documented.

For global level assessments a simple approach is to refer to the numbers of disaster deaths and to those estimated affected by disasters, such as from data provided by EM-DAT, reported by IFRC according to country development categories of the World Bank or UNDP. This information can be found in the World Disasters Reports of IFRC based on the Human Development Index (HDI) of UNDP, which we described in Chapter 2. For example, using 2001 data from IFRC, Wisner *et al.* (2004) show an inverse relationship between deaths and financial loss impacts in relation to country levels of human development. Overall, the low human development category was associated with 1,052 deaths per disaster, whilst the medium and high development categories were associated with 145 and 23 deaths per disaster, respectively. Financial loss per disaster was less in the low human development category, at US$ 79, but US$ 209 and US$ 636 in the medium and high human development categories. Table 3.1 shows a related but somewhat different representation based on the same HDI country groupings. In this instance, impacts are shown relative to different population levels. This also shows that the death rate is significantly greater in the low HDI part of the world, but that relative to population size it is only slightly higher in the medium HDI than the high HDI category. However, there is a significantly greater number affected by disasters in the medium HDI category than the low HDI category. The number of people affected in the high HDI category is very low in comparison to both of these. The estimated damage quoted in US is significantly greater in the high HDI category than the medium HDI category, which in turn is significantly higher than in the low HDI category. However, if we calculate US damage relative to overall population levels we find little difference in per capita finance impacts between the medium and low HDI categories. We can consider this as the 'Asia effect', where countries with a relatively steep development trajectory are nonetheless still experiencing the severe impacts of environmental events.

Table 3.1, shows disaster deaths, affected people and economic damage rates for each HDI category. A much greater number of people are

Table 3.1 ***Estimated economic damage of disaster, compared to disaster death rate and people affected, for three human development levels***

	Reported disasters 1994–2003	*Population 2004 (billions)*	*Deaths 1994–2003*		*Affected people 1994–2003*		*Economic damage 1994–2003*	
			Thousands of people	*Rate per 100,000 people*	*Millions of people*	*Rate per 1,000 people*	*US$ in billions*	*US$ per capita equivalent*
High HDI	1,327	1.3	59	5	42	33	422	331
Medium HDI	3,254	4.4	286	6	2,369	534	239	54
Low HDI	1,096	0.6	328	57	172	300	30	53
All	5,677	6.3	673	11	2,583	404	691	108

Note: HDI is the Human Development Index of UNDP, as explained in the text. Data are extrapolated from a combination of UNDP 2007 and IFRC 2004.

affected, relative to death rates, in the middle HDI (calculated at 890 affected to every 1 death) than in the high HDI (7 affected for every 1 death), and the low HDI (5 affected for every 1 death) over the period 1994–2003. In estimating numbers of people affected, it is possible that some people are counted within this statistic on more than one disaster occasion where disasters are recurrent over this period. However, there is a viable explanation in that middle HDI regions include disaster affected Asian countries. For example, floods in China affected 150 million people in July 2003, and just two floods the same year affected 7.6 million people in India. Although some caution is needed concerning the sourcing of disaster data, in terms of evaluating disaster impacts relative to macro-scale human development, we might safely conclude that:

- disasters have a significantly greater impact in terms of mortality rates in the low HDI category;
- disasters affect (but not necessarily kill) more and greater percentages of people in the medium HDI category;
- disasters have a greater recordable economic impact in the high HDI category;
- the per capita economic damage between low and medium HDI categories is not significantly different largely due to higher population levels in Asian middle HDI countries compared with African low HDI countries;
- financial costs of disasters for the 10 years 1994-2003 are estimated at less than a year of gross national product per capita for the poorest nations–financial impacts are, however, less of a concern where national finance may be already distant from the rural poor, as in much of Africa and some other parts of the world.

This information, though based on rough estimates from more immediately available broad-scale data, is consistent with one of the lessons underpinning this book, that disaster impacts are both a consequence and a cause of underdevelopment. We note here that the full costs are not actually measurable in economic terms.

In addition to loss, disasters also result in opportunities. The old adages that adversity brings out the best in people and that necessity is the mother of invention have been exemplified in local responses to disaster around the world. For example, during the great flood of Mozambique in 2000, those with boats developed a relatively successful business transporting people back and forward between two ends of the severed roads of Gaza province. Some of the economic gains, such as in creation of opportunity through the industry of aid and reconstruction, might also be significant but it is unlikely ever to be possible to fully quantify them. The need to populate such an extensive industry can stimulate additional local employment opportunities. The impact of economic aid in terms of disruption to local livelihoods can, however, counteract this benefit, a theme that has been mentioned elsewhere in this book. That disaster related financial relief can breed corruption, dependency and discontent creates further issues exemplifying how disasters indirectly impact on development to the benefit of just a few people in positions of power.

Human security and development

In this section we consider how ultimately the definition of disaster impact is to do with a loss of human security. The United Nations Office for the Coordination of Humanitarian Affairs (OCHA) Human Security Unit which is representation of this terminology would suggest it includes dealing with the evolution of threats (Commission on Human Security 2003). These are suggested in terms of intra-state conflicts, ethnic confrontations, terrorism, forced displacement, extreme poverty, environmental degradation and infectious diseases. However, the UNDP in 1994 tried to bring the concept away from narrow definitions based on the nation-state more towards the state of individuals. The United Nations University has set up an entire Institute for Environment and Human Security (UNU-EHS) at Bonn. The more detailed definitions imply its importance to development status, including links between social, political, economic and environmental conditions, but fundamentally linked to human freedom and fulfilment. The former Secretary-General of the United Nations, Kofi Annan, said that 'Human Security can no longer be understood in purely military terms. Rather, it must encompass economic development, social justice, environmental protection, democratization, disarmament, and respect for human rights and the rule of law' (Annan 2005). A macro, external or structural view of security in the world is clearly crucial, but the understanding of security includes the role of individuals and communities to place it within a development framework. Kaldor (2007) argues for combining both the emphasis on security in the face of political violence, and development as a security strategy.

A more localised and individualised assessment of human security is typically dependent on individual, household or local community primary subsistence needs, namely sufficient food and water, shelter, an energy supply and safe environment. The UNDP (1994) flagged it as involving seven core elements of economic security, food security, health security, environmental security, personal security, political security and community security. Much of this security is made available in a context of sufficient and sustainable livelihoods, social support, human rights and representation. It is an approach to security that starts with people's status of wellbeing. Donini *et al.* (2005) point out that perceptions of what constitutes security often vary from those of assistance agencies, to include quality-of-life elements beyond physical security. Therefore, at its simplest we might consider that where wellbeing is lost insecurity sets in. Conflicts and global insecurity are in turn amplified by processes of loss of a localised sense of security, both materially and mentally. We have learnt the hard way that even an unresolved tension over cultural or religious difference can result in an enormous breakdown of trust and security between cultures. The causes of insecurity are therefore material in terms of economic marginalisation but psychological in terms of a sense of difference from others. One way of considering the more immediate impacts of disaster on human security is through the assessment of loss of a range of localised assets. As such, human security may in part be examined further through the sustainable livelihoods approach, as introduced in Chapter 2. In this instance, balanced assets are maintained in a context of human security (Figure 3.1a). Severely reduced or imbalanced assets (Figure 3.1b) represent situations where insecurity prevails.

Five of the livelihood assets included are those from the sustainable livelihood framework used by DFID (1999) referred to in Chapter 2 (pp 77–78). A sixth asset, political capital, has been added by several commentators, since in a more detailed livelihoods framework by Collinson (2003). Box 3.1 provides a brief description of each asset. Figure 3.1a represents the circumstance where a strong, integrated and balanced asset situation (iii) on the individual, community or wider scale reduces the impact of external shocks (i). The immediate environment (ii) that offsets an impact on this integrated livelihood security comprises increasing capacity, rights, representation, access to resources, empowerment and relative wealth. However, in the circumstances surrounding the collapse of human security represented by Figure 3.1b the security context is overbearing, and in a proximate environment of increasing displacement, abuse, being denied access to resources, loss of representation, vulnerability and poverty, we find livelihood assets reduced and unbalanced. Meanwhile the collapse of a balanced asset endowment or entitlement at (iii) may provide the opportunity for the

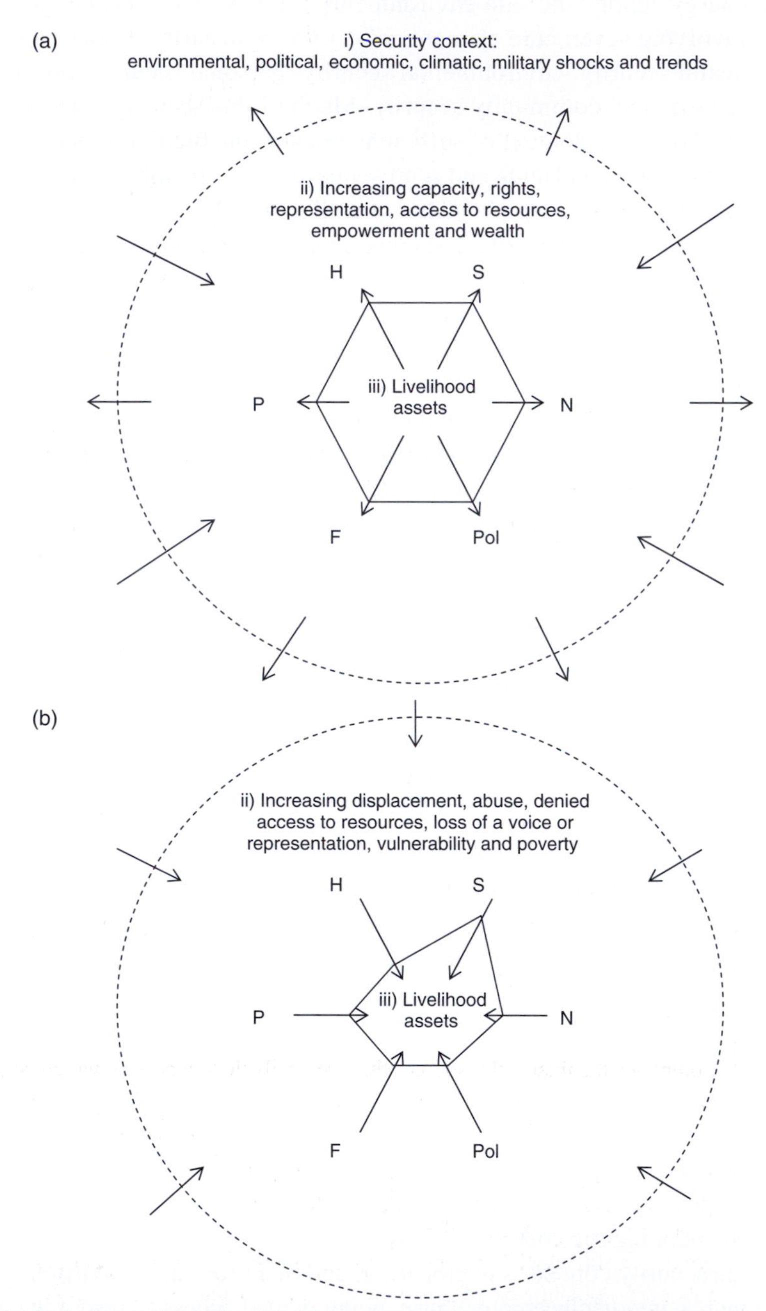

Figure 3.1 Security and livelihoods: (a) increasing human security, (b) diminishing human security, where H = human capital, S = social capital, N = natural capital, Pol = political capital, F = financial capital and P = physical capital

situation at (ii) to expand. Improving the situation arguably requires action at both ends of the system, to strengthen livelihood assets and to reduce the pressure of the wider security context. However, one of the criticisms of development assistance and humanitarian aid might be where it may attempt to apply intervention in an ineffective way at (ii), without adequately understanding the full balance of circumstances at (iii), or reducing the weight of circumstances at (i).

Box 3.1

Capital assets that contribute to human security

Human capital

This is the skills, knowledge, health, ability to work, all of which are needed to make use of the other assets and to build security. A development of specific health security is provided in Chapter 4.

Social capital

This is the family, clan neighbourhood, religious or association networks and connectedness that strengthen social groups and relationships of trust needed to form a safety net and achieve positive livelihood and security outcomes.

Physical capital

Comprises the infrastructure, goods and services–for example roads, water and electricity supply, shelter, information technology, emergency services, engineering equipment–that enhance livelihoods and security.

Natural capital

Includes the land, water (sea, fresh water, rainfall, etc.), soil, forests, air quality needed to achieve livelihood and environmental security.

Financial capital

Comprises the financial resources that are available to a person, which could include savings, credits, remittances and pensions, and which are more secure if from varied sources.

Political capital

This is the presence of different scales of personal rights, including gender equality, political representation and regular access to vote on who has power and how the powers that be influence one's livelihood and security. This asset was absent from some livelihood representations.

Source: Based on Chambers and Conway (1992), DFID (1999), Neefhes (2000) and Collinson (2003).

In terms of a time-frame, changes in this model might be seasonal, the result of a sudden catastrophic impact, encroaching underlying trends or combinations of these. It is important to note that the use of this type of model emphasises not just the magnitude of the threat, but also security in terms of the level of individual and community organisation, coordination and overall capacity to strengthen assets. As such, it is in part a socio-behavioural approach to understanding disaster and development security and resilience.

Beyond the balance of capital assets, insecurity might more simply be considered as being caused by poverty, greed, exclusion and conflicts of culture. This is consistent with a more political economic interpretation of disaster and development resilience and security. Whilst more consistent with radical interpretations, in many ways insecurity through poverty is key to understanding the assets approach to security presented here. It is important to note that building security relies on addressing the causes of real and perceived insecurity, rather than controlling responses to insecurity alone. Human security lies at the intersection of the human, environmental and technological systems necessary for human survival and wellbeing. Evidence is documented in various parts of the world for the cases of health, environmental, food, and livelihood security. This human security is therefore core to (sustainable) development, a lack of it ultimately increasing disaster risks. Given its centrality to human wellbeing, security is a basic human need and right. Early warning, risk management and emergency preparedness can help bring security and are many times more cost effective than responding once security, or a sense of it, has broken down.

Building resilience to insecurity involves multiple stakeholder interests in human security, both real and perceived. However, people's different notions of the causes of and reactions to security risks, particularly in the urban environment, are as yet not fully understood. However, security in the context of disasters is also inevitably represented in military and defence terms. The Civil Contingencies Act (2004) and the consequent emergence of UK resilience forums may have been in part a response to flood, disease epidemic and fuel crises in the UK, but quickly also became the forum for inter-sectoral preparedness for terrorism. This came about through pressure on government to address attacks on security following the bombing of the World Trade Center in 2001 and subsequent related threats such as the bombing of the London transport system in 2005. They have also been in part a response to a lack of local disaster preparedness structures and in particular a lack of community embedded disaster risk reduction. However, much of the emergency management literature of recent years relating to contexts such as the UK has highlighted that there is still a long way to go in providing structures that are felt to be part of the community. The emergency services are based on a hierarchical command and control structure within which decisions are ordered from the most senior downwards. This has been seen as the most effective way of maintaining order and efficiency (Home Office 1994; Alexander 2000; O'Brien and Read 2005), and consequently security. Understanding different notions of security for more secure environments also requires addressing what makes people feel vulnerable in relation to different real and perceived risks.

Coping, wellbeing and resilience

The impact of disasters on development is dependent on the extent to which people, institutions and life support systems can cope in adversity. In many ways the ability to cope underlines what we mean by resilience in disaster management. An emphasis on resilience rather than just disaster response and recovery has become a mainstream idea in disaster reduction. The rationale of being more resilient as a form of disaster prevention is a part of what lies within much of the Hyogo Framework for Disaster Reduction launched in 2005, and many of the national disaster plans of our times (see Box 1.3, p. 43). Coping is an activity that can be considered as old as civilisation itself. To cope in adversity is after all a natural human instinct. Its importance in disaster mitigation, response and recovery is determined also by the extent to which we can use the understanding and reapplication of coping as a disaster reduction strategy, rather than ad hoc responses to crises. Some definitions of coping follow:

> The ability to cope with threats includes the ability to absorb impacts by guarding against or adapting to them. It also includes provisions made in advance to pay for potential damages, for instance by mobilising insurance repayments, savings or contingency reserves.
>
> (UNEP 2002: 426.)

> The means by which people or organisations use available resources and abilities to face adverse consequences that could lead to a disaster. In general, this involves managing resources, both in normal times as well as during crises or adverse conditions. The strengthening of coping capacities usually builds resilience to withstand the effects of natural and human-induced hazards.
>
> (UNISDR: 430.)

One of the limitations of the concept of coping with disaster can be that it does not imply a very developmental approach if seen as just about survival and minimum standards for getting by in adversity. The issue here is that those coping with a crisis aspire to do much more than just get by. They would prefer to achieve a standard of wellbeing and become out of reach of most disaster possibilities. Resilience might be interpreted as also partly defensive, but is distinctly more creative than merely seeking to identify vulnerabilities. It would imply being strong, with the means to cope, rather than lacking something. However, many definitions have been used in recent years as this is a concept with a variety of disciplinary origins, all with different dimensions to its application (Manyena 2006), from technical, ecological, sociological, economic and infrastructural, and organisational versions, to name just a few. The principles of resilience in disaster reduction are multiple, but include robustness, resourcefulness and proactiveness, rather than just making do with minimum standards. The following is a useful definition of resilience in terms of disaster reduction:

> The capacity of a system, community or society potentially exposed to hazards to adapt, by resisting or changing in order to reach and maintain an acceptable level of functioning and structure. This is determined by the degree to which the social system is capable of organizing itself to increase its capacity for learning from past disasters for better future protection and to improve risk reduction measures.
>
> (UNISDR)

Being a mainstay of current thinking and policy development in disaster reduction, resilience also featured as the core theme of the 2004 *World Disasters Report.* An entire field of integrated disaster studies is now dedicated to disaster resilience as the adaptation that is required (Paton and Johnston 2006). This is also being helped along by the Climate Change agenda of the IPCC, which, beyond assessing change in climate risks, advocates assessment of human adaptation, coping and resilience to its effects.

The key point about the resilience approach is that whilst insecurity is generally bad for development, under pressure people cope and become resilient, reducing disaster impacts in the process. Local risk assessment studies have been known to contribute to identifying what really makes a community more resilient. Whilst vulnerability assessment focuses on the deficit in people's lives, assessment of resilience indicates what can be positively enhanced. This activity is returned to in Chapter 5. The gatekeepers in determining the extent to which the community can build resilience are both public and private institutions that influence people's sense of security. The risk governance of human habitation presents a challenge in that risks are concentrated in increasingly high densities in cities. Meanwhile, coping with risks is an issue driven by representations in the media. However, risk governance may be improved by knowing to what extent different citizens feel a sense of insecurity. A sense of insecurity may be from more local hazards, such as being attacked in the street, or from more distant threats, such as, for example, global conflict, economic collapse, pandemics or climate change. Knowing the manner in which people interpret, react to and develop resilience to risks more locally can contribute to security by shrinking the space between knowledge, communication and action for risk reduction. Some examples can be observed from initiatives in the developing world, where the pressure to come up with individual and community based approaches, often in the absence of any other alternative, has been most acute.

Plate 3.2 Cyclone shelter in coastal Bangladesh that was donated by the Saudi Arabian Government. This is used as a school, mosque, and community centre when there are no cyclones. Its combined functions go some way to exemplifying what might be meant by infrastructual and developmental resilience.

Source: Author.

For simplicity, two broad categories of resilience can be posited, one that emphasises something static and solid, and the other a reliable and secure process. The former is exemplified by infrastructure such as sea defences, earthquake resistant buildings, cyclone shelters and the like. However, resilient processes in disaster management comprise the systems of organisations and procedures that make them function. As hazards and susceptibilities constantly change, a resilient process is one that is capable of ongoing adaptation. For example, should communications break down between the emergency services during disaster, their resilience is defined in the strength of the contingency plan needed to restore them to functioning again. Managing the influence of disaster on developed resilience therefore goes beyond individual strength and engages early warning systems, community preparedness and institutional contingency development for coping with disaster. This is one of the key concerns of the field of business continuity management (BCM), which seeks to protect businesses from experiencing catastrophic breakdown during emergencies. In particular, the coping and resilience of businesses in the age of information technology is dependent on being able to protect critical infrastructures (Hyslop 2007).

The latter has in recent years also been a development within UK disaster preparedness, including through the establishment of the Regional and Local Resilience Forums. Furthermore, it is a concept of developing and post-conflict areas. Several projects engaged by the author in Mozambique, Nepal and Bangladesh are experimenting with local risk and resilience committees, owned by local community structures, the purpose of which is to identify and then intervene in local risks. The themes of resilience and security underpin much of the disaster and development debate and are returned to in the later part of this book.

Population displacement and refugeedom

Disasters dislocate people from their normal ways of life. Those who survive disasters may do so with loss of home, family, community members and livelihood. Displacements of people can be both an indicator and cause of underdevelopment. To examine this process further, we may consider different categories of population displacement, the displacement process or cycle, and the impact of displacement on local economies, environments, society and health.

Displaced people come to be referred to as 'refugees' when their forced relocation from one place to another becomes institutionally recognised.

After the mass displacements of the Second World War there was renewed motivation to define this status. The United Nations set up its High Commission for Refugees (UNHCR), the criteria being the following:

> Those forced to leave their homes as a result of a well-founded fear of persecution for reasons of race, religion, ethnicity, membership of a particular social group or political opinion.
>
> (1951 Refugee Convention, Geneva)

The 1969 Organisation of African Unity Refugee Convention and the 1984 Cartagena Declaration in Latin America expanded the definition to include people who have fled because of war or civil conflict. A total of 146 countries have signed the 1951 UN Refugee Convention and/or its 1967 Protocol recognising people as refugees based on these definitions.

Traditionally, the designation of refugee meant that you had also to have crossed an international boundary. This meant countries needed to have asylum regulations to decide who is and who is not a refugee. 'Asylum Seeker' is the legal term for a person who has arrived in a country who is seeking a positive decision on their asylum claim. In the case of the UK this would be whether they have been granted 'Leave to Remain' or 'Protection'. The role of the UNHCR has been to try to uphold the international dimensions of that process and of the types of humanitarian assistance that may be required in more extreme mass displacements. The 1951 definition implies refugee status based on basic human rights. The Universal Declaration of Human Rights itself states that everyone has the right to seek and to enjoy asylum in other countries as a means of protection from persecution.

Harrell-Bond defines refugees as

> a particular category of people who, because they have lost the protection of their own state, must rely on the willingness of others to observe humanitarian norms, some of which have been codified in international law.
>
> (Harrell-B. and *et al.* 1992 : 206)

But defining who is and who is not a refugee, let alone defining displacement, is far from simple. For example, Zetter (2007) revisits the concept of refugee labelling, suggesting that 'convenient images' of refugees and refugee discourse are driven by the need to manage globalised processes and patterns of migration and forced migration. Forced migrants include those who are institutionally labelled as refugees and internally displaced people (IDP) (Norwegian Refugee Council 2002), amongst whom may be included those displaced by flood, drought, famine, desertification, collapsed economies and hostilities. Displaced people may be political refugees, variously escapees, evacuees, expellees, all of whom may be forced to resettle within or without their own country boundaries. The UN in 1998 established 'Guiding Principles on Internal Displacement' (Norwegian Refugee Council, 2002:191) that in effect notionally give rights to displaced people who have not left their country, thus getting over some of the problems of the dilemma of classifying refugeedom. However, as displacement is a continuum of complicated conditions, the UNHCR often now simply refers to 'those of concern to UNHCR' as its intended beneficiary target group. At the start of 2006, the number of people 'of concern' rose to 20.8 million, up 6 per cent from the 2005 total of 19.5 million (UNHCR 2006). The increase reflected change amongst groups, including refugees, civilians returned home but still needing help, internally displaced people, asylum seekers and stateless people.

The closure of boarders to prevent unwanted refugee influxes, such as occurred between Kosovo and Macedonia, also contributes to a relative increase in the ratio of refugees to IDPs. There were more than 23 million IDPs by 2000, compared to 14 million refugees in 1995. Principles of the right not to be repatriated have also been frequently overridden. One example was the repatriation of 1.2 million Rwandan refugees from Zaire and Tanzania in late 1996. Prior to this, one of the big issues that was brought to a head concerning refugee identity was over the case of humanitarian assistance to Rwandan refugees whilst in Zaire in 1994. The issue was about how best to help the 1.2 million refugees when 5–10 per cent of them were not *bona fide* refugees, but associated with the Hutu militia that committed the genocide that caused the refugee situation in the first place. Eighteen NGOs called for a force of 5,000 UN troops to secure

Table 3.2 ***Persons of concern to UNHCR by region, 2006***

Region	*Number at 1 January 2006*
Asia	8,603,600
Africa	5,169,300
Europe	3,666,700
Latin America and Caribbean	2,513,000
North America	716,800
Oceania	82,500
Total	20,751,900

Source: UNCHR (2006).

the camps around Goma and Bukavu but were ignored. Two NGOs pulled out, but others stayed, accepting that they would just have to provide assistance to everyone regardless of whether they were genuine refugees or not. Similarly, Pol Pot's Khmer Rouge, responsible for the deaths of anything up to 3 million Cambodians, an estimated 1.7 million of whom where directly murdered by his troops during the period 1975–1979, subsequently mixed with refugee groups fleeing the Vietnamese. These fighters were subsequently indirectly assisted by the UN on the eastern border of Thailand. It could be argued that militias are refugees also. These incidents have raised important issues about the definition of the refugee.

In terms of the impact of disasters involving population displacement and of the impact of displacement on development, a simplified version of three broad categories may be distinguished. These include refugees on whom the status of refugee has been conferred, those internally displaced or without necessarily gaining refugee status, and those who are not displaced from one location to the next, but displaced *in situ.* The third category is defined through the destruction of the immediate surrounding environment on which people are dependent. The displacement of people through the dislocation of their relationship with their immediate surroundings may be a form of 'topocide', literally the destruction of places, where places comprise the social, economic and environmental fabric of survival. The terminology has also been used by Porteous (1989) in describing a place being 'planned to death' through development in Canada, meaning that local identity and continuity with the past had been lost in new designs.

Extreme cases of topocide are represented by the case of the destruction, at the end of the Second World War, of Dresden, a place that was formerly the cultural capital of Europe. There are many examples, but a further extreme case was that of Pol Pot's Khmer Rouge during the early late 1970s and 1980s setting about creating 'year 0' for Cambodia. About 3 million people were killed, and most of the country's population displaced in the process of an extreme Maoist interpretation of asset redistribution and re-education, until Vietnamese troops intervened. Other examples of place destruction, often less brutal but devastating in the longer run, are the millions of people who have been uprooted from home and livelihood by dam construction. This form of development induced displacement has been responsible for the displacement of millions in China, India, South America and Africa. National parks and mining have displaced a further cohort of millions. There is fairly well-established literature on these topics, a good review of it being provided by Cernea (2005).

The mixed condition of displacement is one that refugee scholars recognise only too well. It extends also to separation from a psychological placement. Gebre and Ohta (2005), in acknowledging the complexity of displacement studies, emphasise the need for a holistic and integrative approach in addressing displacement risks in Africa. Whilst this is a sound approach it is important to emphasise throughout that displacement for most is a form of destitution. Simply put, this is according to the *Oxford English Dictionary* a 'state of being extremely poor and lacking the means to provide for oneself'. The condition can be extended to the lack of any reasonable statutory support mechanisms. Cernea (2005) has included destitution as part of impoverishment in his IRR model. This is built on three concepts: risk impoverishment, and reconstruction. As a generic model it has been used in displacement studies to assess potential risks through displacement against the actual outcomes in different contexts. The eight most common impoverishment risks in the model are:

1. landlessness;
2. joblessness;
3. homelessness;
4. marginalisation;
5. increased morbidity and mortality;
6. food insecurity;
7. loss of access to common resources and services;
8. social disarticulation (community dis-organisation).

(Cernea 1997, 2000, 2005)

Additionally, there are economic migrants, those considered not to have been forcibly displaced but who have migrated voluntarily on account of strong pull factors towards the lure of greater prosperity. However, the designation of economic migrant is once again an unclear one, as it is also arguably the push of poverty in one's home area that drives people away, therefore also a form of forced displacement. This may become a clearer driving force if climate change predictions are correct. The IPCC (2007) has indicated that by 2080 1.1–3.2 billion people will be experiencing water scarcity, 200–600 million hunger and 2–7 million coastal flooding. This has led to some estimations of an additional 250 million people being permanently displaced by climate change phenomena alone by 2050. Further to this, some new categories of forced migrants may be identified in the context of worsening health conditions, such as have already been realised for the case of those migrating in search of medical attention, including HIV/AIDS treatments.

Some idea of the figures available for various types of forced migrants (but omitting those only very locally uprooted) can be obtained from the

International Organisation of Migration (IOM), another institution that is part of the UN family of organisations. Recent trends for refugeedom and IDPs are available from the Global IDP Survey, available from the Norwegian Refugee Council. However, most data tend to account for one or other type of displacement category. For an overall estimate, a summary by Christian Aid has been included here (Table 3.3). This organisation has interpreted information from several sources to be able to very approximately quantify numbers, but add that in their view these are conservative estimates of the reality of the situation by 2007 and projection until 2050. The projection includes those that would be displaced should current predictions of climate induced disasters be taken into account.

Disasters do not impact on people evenly and consequently those most displaced are likely to be the most vulnerable children. Particularly vulnerable groups during mass displacements are often under 5-year-old children, orphans, pregnant/perinatal women and those nursing small children, those without any family, the old and dying. However, this is highly dependent on the nature of the displacement. Thus in the Horn of Africa the death toll among refugee groups has been highest among the youngest when the main hazard has been malnutrition. However, during

Table 3.3 ***Overall estimates of global population displacement***

	Forcibly displaced as of 2007	*Projected forcibly displaced – present to 2050 assuming continuation of current trends*
Conflict and extreme human rights abuses remaining in own country	25 million	50 million (i.e. one million per year)
Environmental disasters (earthquakes, hurricanes, floods, etc.) remaining in own country	25 million	50 million (i.e. one million per year)
'Development' projects such as dams, mines, roads, factories, plantations and wildlife reserves, the vast majority remaining within their own countries	105 million	645 million (i.e. 15 million per year)
Refugees, those fleeing persecution in their own countries and gone to other countries that have accepted their claims for asylum	8.5 million	5 million
Permanently displaced by climate change-related phenomena such as floods, droughts, famines and hurricanes	Not quantified	Additional 250 million
Total	163 million	1 billion

Source: Christian Aid (2007) (based primarily on data from UNCHR).

Plate 3.3 Women on the move in Western Darfur, Sudan
Source: Janaka Jayawickrama, Northumbria University.

the conflicts in Darfur, Sudan and the Balkans adult male refugees were more at risk due to being potential combatants and therefore specifically targeted by opposition fighting forces.

In order to understand the impact of population displacement on human wellbeing, it is useful to consider the variety of risks encountered throughout the process of displacement and how these change over the displacement process. People become more vulnerable and are exposed to new circumstances of environmental, social, political and economic hazards. Table 3.4 outlines the different displacement factors and individual variables that can impact on people through the process of uprooting to resettlement, a process that has otherwise been referred to as the 'refugee cycle'. The displacement (or refugee) process or cycle separates out different stages, each of which has an impact on the wellbeing of the individual or community affected. Numbers of displaced people have been large enough for this to be an influence on development for significant proportions of both the displaced and host communities. For example, during the late 1980s it was estimated that in addition to

Table 3.4 ***Influences on the condition of displacement***

Displacement factor	*Variables*
Cause of the initial displacement	War; famine; disease; environmental disaster (i.e. floods, droughts, earthquakes, landslides, and environmental degradation); development projects (i.e. dam construction), economic decline or restructuring; political repression; ethnic cleansing; societal decay and crime
Characteristics of the place from which displaced	Rural; urban; livelihood system; relationship to local environment (i.e. in terms of use of water, sanitation, cultivation system and fuel supply); ethnicity; sense of community; historical and cultural value
Manner of flight at time of displacement	Rapid or gradual; level of force; social composition of those relocating; choice of staying behind; mobility of assets
The displacement track	Duration and distance; mode of travel; diffusion pattern (i.e. direction and contiguity of small or large groups of migrants, or individuals); degree of harassment or support; mobility of assets
Destination	Rural areas • Vegetated, wet or dry conditions, productive or non-productive land, low lying or mountainous Urban areas • Peri-urban, or inner city, ghetto or dispersed Refugee camps or displacement centres • Open or closed access, density of habitation, ethnic, political or religious composition of overall group of occupants General • Type of aid and assistance provided, upholding or not of rights and responsibilities (i.e. security, the Humanitarian Charter, and the right to seek employment and build a livelihood)
Circumstances of settlement in host area	Behavioural adaptation of the displaced; receptiveness of the host community; integration and community cohesion; maintenance of links with origins; environmental resilience; capacity of infrastructure Capacity of existing infrastructure; institutional, community and individual observance of displaced people's rights (i.e. security, right to work); observance of the Humanitarian Charter
Re-displacement	Onward migration that is forced or by choice; within same country or to a third or fourth country; changing status of refugeedom at different hosting areas
Return migration	Persistent links with place of original displacement; assisted or non-assisted; legally accepted or illegal; selective return dependent on social group; rate of return; relocation to original, similar or totally new environment; maintenance of links with place of refuge; re-occupancy of own dwelling, occupancy of newly created dwelling, or forced displacement of a further group of occupants at place of return; rate of reconstruction and recovery

over three million refugees from Mozambique fleeing conflict, a further six or so million were internally displaced or camped next to the ruins of their villages (Green 1992). In all, therefore, over half of the population of the country was displaced by war and accompanying famine.

This phenomenon in human history is alarmingly frequent. By August 2007 it was estimated that over two million Iraqis had been displaced from their place of residence since the start of the upheaval brought by the US invasion of 2003. Circumstances of displacement influence not only the roles of the displaced and host populations, but also humanitarian response interventions, longer-term integration and/or resettlement.

Although much has been made of the negative development impacts of displacement, it is important to bear in mind that refugee movements around the world have also demonstrated how refugees and other displaced people can end up being very successful. There is a long history of instances of refugees having stimulated local economies in the process of rebuilding their own livelihood security. It is therefore sometimes considered a missed opportunity that resettlement and refugee policies are frequently not conducive to letting people rebuild their lives. The refugee policies environment can therefore constrain development. Both positive and negative impacts, somewhat complicated and contradictory, are identified by Ohta (2005) for the case of the hosting Turkana and refugees in northwestern Kenya. Given the right policy context and opportunity, refugees' own social networks and coping mechanisms have been found to be the method through which refugees have been able to rebuild their livelihoods post-displacement (Willems 2005). This perspective raises questions about the usefulness of controlling the movements and activities of displaced people at all.

Other studies of population impacts have considered this point in looking at the impact of displaced people on the environment. For example, pressure on natural resources around refugee reception areas, and the political ecology of combined refugee and environmental issues are addressed by Black (1994, 1998). Whereas mass migrations have an environmental impact on hosting areas, following resettlement those areas have been found to regenerate once refugees move on or are allowed to use the environment for sustainable livelihood development. During periods of pressure on natural resources, however, refugee groups may enter into tensions with host communities dependent on the same resources, as identified by Mbakem-Anu (2007) for the case of mixed origin refugee groups subsisting in Congo–Brazaville. This was also indicated by Kurimoto (2005) in an extreme case from western Ethiopia.

One of the common features in this tension has often been a lack of foresight and competent planning by governments and aid agencies who may have been able to influence the direction of displacement flow and resettlement. In other instances resettlement has arguably been used as a political tool to manipulate international aid, ethnic allegiances and political alignments, as claimed for the case of Ethiopia in the 1980s. Gebre (2005) records how the Ethiopian Government relocated 600,000 people from drought affected and overpopulated regions of the country to five sites in the northwestern, southwestern and western parts of the country, 82,000 of whom moved into an area of shifting cultivation. The official explanation for this resettlement, which some claim was forced, was to avoid famine and dependence on food aid. The international community did not intervene. The real reason is thought to be that the government of the day was trying to suppress insurgency and dilute ethnic homogeneity in the areas from which the people were removed. Coupled with an ineffective international relief and development programme in the new areas, the displaced people found little recovery possible. Many are known to have starved to death post-resettlement.

Health impacts of displacement

Health impacts of displacement have also been a key concern of disaster and development studies. With increases in ill health, recovery and development are held back. There are psychological impacts from displacement as well as the physical health impacts. Displaced people may become primary targets of ethnic cleansing, murder, sexual violence, torture, and mutilation, often with the purpose of demoralising or annihilating a community and its identity. For refugees, particularly in Africa, this has often been part of a protracted set of risks that are ongoing throughout various stages after initial displacement (Crisp 2005). Across these examples, the departure from the norms of human behaviour constitutes the basis of what has been classified as 'trauma'.

Conflict induced mass displacement, such as experienced throughout the twentieth century, including during two world wars, has major consequences for public health and the longer-term wellbeing of communities. The resultant deviation from the norms of life disrupts not only social and psychological sound-mindedness but also socio-economic development, reconciliation and peace. Whilst health responses understandably put an emphasis on nutrition, prevention and control of infectious diseases, maternal and child health, there is an often neglected demand for recognition of the importance of restoring mental health during and post emergency. Earliest material responses to needs, with provision for food, water, shelter and protection, are, however, arguably also a preventative mental health response. The question is what do survivors of disasters or conflict request for health or psycho-social support? It is argued that from the 1940s until the end of the 1970s humanitarian dialogue tended to engage with displaced or disaster and violence affected communities as resilient and strong, whereas communities more recently have been seen as vulnerable and weak (Jones 2004; Argenti-Pillen 2003; Bracken *et al*. 1995) This perhaps reflects the rise of a supply driven culture within some aspects of the relief and development industry.

Resettlement resulting from conflict, unrest or environmental catastrophe has increasingly become associated with communicable disease outbreaks (Shears and Lusty 1987; Toole and Waldman 1993). With increases in the number of recognised refugees and internally displaced civilians in the world during the early 1990s to 50 million, the association between population displacement and health re-emerged became reawakened as an increasingly important issue (Toole 1995). Also, what were thought to be old diseases of underdevelopment were clearly not on the wane, as had been predicted in the 1950s. In these circumstances, congestion of people into disease conducive environments, often combined with increased human vulnerability through having been forcibly displaced, can accompany negative stresses on local disease ecologies. One way in which this can occur is when large numbers of internally displaced populations (figures are unknown) seek refuge in urban areas.

Displacement also sometimes causes the creation of new urban areas that are not based on pre-existing settlements. However, based on multiple sites in Mozambique, Collins (1998) found that higher incidence of specific diseases amongst displaced people in comparison to existing residents living in the same areas, depended on the local environment and the primary needs of the hosts. In these circumstances the incidence of disease among those ‘displaced’ was not always very different from general standards of health in those areas. Whether or not a person was in the category of having been displaced was sometimes less important than where each person (displaced or otherwise) currently had to reside. The full explanation of disease distributions seemed to suggest that, whereas population displacement invokes change born out of forced mobility, modifications to infrastructure and policy may similarly ‘displace’ the wellbeing of people in their place of residence. Therefore, the identification of geographically defined risk areas and processes of change, such as those associated with resettlement and being displaced, together with the governance of the areas, is crucial to environmental health management.

Through displacement there can be changes in environmental location experienced by displaced people that expose them to increased disease hazards and/or make them more susceptible to health problems. There can also be impacts by displaced people on local environments that create more favourable conditions for disease. Also, behavioural differences between displaced and host communities and behavioural changes associated with the prevailing conditions at different resettlement locations may have an influence on health outcomes. Finally, the environmental context of settlement and disease, as with other health and environment issues, can be viewed with respect to wider issues of human development.

Emergency relief and development

One of the legacies of humanitarian assistance has been that it is the necessary emergency reaction required to deal with disasters that have already begun to take their toll of human life and suffering. It is the curative part of disaster management either once a crisis is underway or after its main impact has already occurred. Though in terms of cost-effectiveness investment in disaster prevention would vastly save on the cost of humanitarian assistance (Linnerooth-Bey *et al.* 2005), millions of lives have been saved through emergency relief. It is what is required when no other means of dealing with crisis is available. Well-targeted relief saves lives and is withdrawn once its job is done. However, relief can also create dependencies. Its impact on development is positive inasmuch as it prevents loss of life during emergency, but can have some negative effects if local systems and the means to rebuild economically are undermined by it. For example, provision of emergency food aid in a region where food is normally produced will bring down prices, potentially destroying local markets. In some of the poorest parts of southern Africa, provision of second-hand clothing from Europe has been known to have forced local producers to stop manufacturing and trading in local products.

Several of the relief agencies have recognised the limitations of emergency relief and also engage in development work. The move from relief to development is now a common objective, not only in terms of more preventative approaches to disasters and to longer-term recovery, but also so that agencies can remain active during non-disaster periods.

Imaging, imagining and representation during disaster

Emergency relief and development work have variously been limited in recognising what is meant by local coping and resilience. This exposes a deeper problem concerning the categorisation of disaster survivors as helpless victims, rather than capable people. In responding to disaster issues, imaging, imagining and representation of a 'disaster' by the media have had a significant influence. This has a direct bearing on when and how aid ends up being delivered, with consequences for development in the recipient zones.

Smith and Dowell (2000) identify coordination difficulties in emergency management as being limited by poorly shared mental models and 'a possible conflict between the requirements of distributed decision-making and the nature of individual decision-making' (2000: 1153). Cox *et al.*, in an analysis of local print-news media coverage of the recovery process in two rural communities following a devastating forest fire revealed a 'neoliberal discursive framing of recovery, emphasizing the economic-material aspects of the process and a reliance on experts'(2008: 469). They found that the 'dominant "voice" was male, authoritative, and institutionalized' (2008: 469). In a study of the reporting of Hurricane Katrina in 2005, Barnes *et al.* found that the media framed most stories by 'emphasizing government response and less often addressing individuals' and communities' level of preparedness or responsibility' (2008: 604). Some observations on the representation of the recent cyclone in Burma and earthquake in China are presented in Box 3.2.

In the case of famine emergencies in neglected parts of Africa, it has often been the media that have drawn attention to the scale of suffering, stimulating an emergency response from both the public and decision makers. A difficulty here, however, is the tendency to wait for the onset of a disaster event of this type before the images of suffering can be captured for an international audience. What is really required for preventative action in these more predictable instances is careful monitoring of food insecurity indicators linked to an early warning and recovery strategy, as discussed in Chapter 6.

Disaster scenarios and international policy regimes

International disaster reduction scenarios might become better coordinated internationally if the sentiments of Millennium Development Goal 8 come to fuller fruition. This is the call to 'develop a global partnership for development'. Where aid and disaster relief are not well directed, it can not only compromise local development but arguably impact on good governance. If more support goes to disaster relief and away from poverty

Box 3.2

Representations of the Burma Cyclone and Chinese Earthquake of 2008

Cyclone Nargis happened in Burma on 2 May, 2008 and the Wenchuan Earthquake of Sichuan Province, China, on 12 May. These were unpredictable events for which there is thought to have been no prior preparedness. There were an estimated 8 million people displaced by Nargis, with 15,000 dead. The earthquake in China displaced an estimated 15 million people, with 70,000 deaths (www.reliefweb.org). The contrast between the two emergencies was partly in the way the governments of these two countries wanted to portray their situation. In Burma there was a slow response from the authorities, whilst international assistance was at first rejected by the military junta there. The authorities tried to present a scenario whereby they would seem to be in control of the situation, largely due to wanting to keep the international community from 'interfering' in their affairs through being able to observe the country more closely. There was a severe cost to civilians, who consequently did not receive the assistance they might have, should the military junta have responded in a manner that was more in the interests of its people.

In the Chinese earthquake disaster the authorities managed a much more rapid response, starting with local bodies and then extending to bringing people in from other parts of the country. The logistics meant that much of the initial response had to also be by the survivors themselves, helping each other get out of the rubble of collapsed urban areas. However, the international community was allowed to report the disaster at close quarters, and the necessary assistance from beyond China was permitted. The emergency was represented as an open space within which people could operate. The representation of this disaster could only be tragic, but it had been important to China to present itself well in this situation. This was not least due to a need to rebalance recent disfavour internationally over Tibet, and in relation to the Olympics that were due to take place in Beijing just a few months later.

reduction, then relief would be counterproductive. One of the contentions is in terms of the point at which different levels of humanitarian assistance and emergency response are the responsibility of international organisations, central government, local government, community or sub-community levels. This is a debate that appears to have pervaded all parts of the world. For example, in the UK since the Civil Contingencies Act in 2004 there has been a move to decentralise emergency response, with responsibilities at local government level and in regional forums. In the US it was the heavy centralisation of FEMA that led to much of the criticism it received for inefficiency in responding to the Hurricane Katrina emergency. Some countries in the developing world have variously argued for designating ministerial level disaster management responsibility to Environment, Home Affairs, Local Development, Health, Finance or other departments. Some nations have created singularly focused disasters ministries and this practice has increased since the setting up of the Hyogo Accord in 2005. The choice of existing ministries within which to mainstream disasters issues not only can be a political manoeuvre for those in power, but also reflects different interpretations and demands on the sector. The somewhat unique choice by Sri Lanka to establish a Ministry for Disaster Management and Human Rights has been mentioned earlier. A similar sentiment is not yet evident at the international level. However, development and relief oriented by the aid donor countries have been directed increasingly to conflict and humanitarian affairs rather than only those policies amounting to structural adjustments of development. The trend in countries like Sri Lanka, Pakistan and Mozambique is to now adopt a shift from the civil defence relief response modality to a risk reduction-cum-disaster preparedness approach.

Conclusion: sustainable development through disaster management

The impact of disasters on development is widely appreciated but not simple to quantify. Environmental, economic and social crises will slow development but not necessarily prevent ongoing development trajectories for long where they are already well established. Human insecurity, whether through loss of the means to support basic needs, or to protect people from external threats, defines much of the challenge that needs to be addressed through a disaster and development agenda. This is particularly clear from the need to address people's loss of a livelihood. However, human development is grounded in an age-old capacity to cope with disasters. As such, people, communities and institutions can become more resilient. We can learn from human coping and resilience as a way of knowing how to strengthen capacity to deal with future threats. The case of displaced people and the changing hazards and vulnerabilities experienced through forced displacement from one location to the other, across borders or without going anywhere, demonstrate multiple stages of coping and resilience. Disaster scenarios are represented, reacted to and institutionalised in different ways around the world, from the more global to more local. However, despite this diversity a common principle for the twenty-first century is that reducing disasters improves development changes.

Beyond this, disaster reduction that builds on people's individual quest for security and wellbeing to a level beyond the minimum required for survival is likely to be the best formula for promoting sustainable development.

Discussion questions

1. How might we estimate the impact of disasters on development?
2. What aspects of development help people to cope with disasters?
3. What are the different ways in which people experience forced displacement?
4. Is humanitarian relief good for development?
5. Discuss why disaster prevention encourages sustainable development.

Further reading

Bankhoff, G., Frerks, G. and Hilhorst, D. (eds 2004) *Mapping Vulnerability: disasters, development and people*, London: Earthscan.

IFRC (2004) *World Disasters Report 2004: Focus on Community Resilience*. Geneva: IFRC.

Ohta, I. and Gebre,Y. D. (2005) *Displacement Risks in Africa: refugees, resettlers and their host population*, Kyoto: Kyoto University Press and Trans Pacific Press.

Smith, K. (2001) *Environmental Hazards: assessing risk and reducing disaster*, third edition, London: Routledge.

Wisner, B., Blaikie, P., Cannon, T. and Davis, I. (2004) *At Risk: natural hazards, people's vulnerability and disasters*, second edition, London: Routledge.

Useful websites

www.interragate.info: Online Country-by-Country Global Hazard Database, an online natural hazards database designed to assist disaster management NGOs, humanitarian agencies and rescue teams responding to natural disasters.

http://ochaonline.un.org/humansecurity: United Nations Office for the Coordination of Humanitarian Affairs–Human Security Unit.

www.cred.be/: the Centre for Research on the Epidemiology of Disaster manages several disasters databases, including the International Disaster Database, which provides data on both global disaster occurrence and impact

www.resalliance.org/: the Resilience Alliance.

www.proventionconsortium.org: ProVention Consortium.

www.forcedmigration.org/: Forced Migration Online (FMO), of the Refugee Studies Programme at Oxford University, provides instant access to a wide variety of online resources dealing with the situation of forced migrants worldwide.

www.lodihpn.org: the humanitarian Practice Network (HPN) of the Overseas Development Institute (ODI) provides a forum for policy-makers, practitioners and others working in the humanitarian sector.

www.gdnoline.org: The Gender and Disasters Network, currently managed in association with the Disaster and Development Centre (DDC).

Physical and mental health in disaster and development

Summary

- Recognising the nature and context of health issues associated with disasters and development is central to improved health and disaster reduction.
- Disease is the greatest disaster category and often a consequence of other disasters or development processes. This may be through rapid onset epidemics or ongoing endemic health burdens.
- HIV/AIDS is a pandemic infectious disease disaster, an example of an emerged combination of pathogenic, social and development oriented health risks.
- Famines are the consequence of the extreme underdevelopment of food and nutrition security.
- Chemical poisoning disasters are generally development induced.
- Mental health requires varied interpretation of what constitutes trauma and wellbeing in disaster and in development.
- Health belief, health promotion and self-care are core aspects of health disaster prevention.
- The mainstreaming of primary health care principles and practice in health care development is exemplary for wider disaster reduction approaches.
- Integrated health management builds health security that can reduce the risk of disasters.

Introducing health, development and disasters

This chapter addresses some of the multiple aspects of health that lie at the heart of disasters and development issues. The reason for its inclusion as a significant component of a book on disaster and development is also that disease epidemics remain the greatest cause of mortality and morbidity in developing areas, far outnumbering other environmental disaster categories. Also, health prevention has a direct resonance with disaster prevention, the two fields closely overlapping in approach. The other reason is that other disasters impact on health (Noji 1997), and improving health reduces disaster.

In terms of a conception of what is meant by 'health', simple reference is made to the often cited WHO definition of

> a state of complete physical, mental and social wellbeing and not merely the absence of disease or infirmity.
>
> (WHO 1948)

The dictionary definition of wellbeing is 'the condition of being contented, healthy, or successful' (*Collins English Dictionary*). Physical, mental and social wellbeing in many ways define development itself and are ideally viewed together. It is well understood by clinicians that mental health can have implications for physical health and vice versa. Meanwhile social wellbeing is both a cause and a consequence of physical and mental health, but modified by wider development processes and disasters. To be able to diagnose some of the individual components of this wellbeing for contexts of disaster and development it is necessary to separate out different components of overall health status. In doing this, the chapter analyses some of the greatest disaster and development issues of our times; namely, infectious and chronic-degenerative disease, malnutrition and mental health. It then moves on to consider different aspects of health care in disaster and development, including primary and emergency health care systems, social care, self-caring, health promotion and the role of all of these as mainstream to disaster reduction more widely. Ultimately, this integrated approach is the foundation to being able to advance health security for preventing health related disasters.

Health aspects of development are well-established realities of past and present civilisations. Underlying associations are reflected in development perspectives of the last few hundred years. One predominant view, along the lines of Malthus in the late eighteenth century, was that disease and ill health are the inevitable way in which populations are kept in check, with only limited options for full prevention. However, the discovery of microbes (or pathogens) by the mid nineteenth century by Robert Koch brought with it a better understanding of the association of sickness with exposure to disease hazards in different living conditions. Ultimately, we have become better sensitised to the idea of diseases of development and those of underdevelopment, together with the possibilities for their prevention.

The demographic transition model indicates that, rather than there being mass die-offs from disease, population growth can eventually normalise through economic development. Reduced mortality throughout many periods of history was further evidence that potential health disasters can to a large extent be managed. Specifically, changes in the development status of a region revealed how overall population surges start with increases in successful births and reduced infant mortality due to better health care and living conditions (Gray 2001). Meanwhile, improved quality of life contributes to a higher average age of death. Under these circumstances of development, a population increase occurs when the age of death rises and more children survive the first few months or years of life. However, improved wellbeing eventually means people have fewer children, so that despite the population living longer there is a stabilisation of population numbers. This simple model of population and its relation to health and health care is used to describe the underlying demographic change in Europe since the eighteenth century. In earlier chapters we addressed how some development theories assumed that this pattern was replicable in the developing world. Furthermore, where having multiple children becomes less popular and old age more prevalent population booms can turn to population bust, with unnaturally low numbers of young in relation to old people. Recent concern about too rapidly declining population growth is reflected in Box 4.1.

The presence or not of infectious diseases is a key influence on the demographic transition model. The decline of infectious disease in some parts of the world with improved environmental health, medicine and other aspects of health care is, however, only a part of the story of health and development. This is because development can impact on health in other ways, most notably through an increase in chronic degenerative health problems, accidents and poisonings, and conflict related morbidity and mortality. The demographic transition is therefore really an epidemiological transition (Philips and Verhasselt 1994), health transition (McMurray and Smith 2001) or risk transition (K.R. Smith 2001). Epidemiology is the study of the distribution and transmission of disease. The transition occurs when infectious diseases such as diarrhoea,

Box 4.1

Demographic ageing

The High-level Meeting on the Regional Review of the Madrid International Plan of Action on Ageing (MIPAA) took place on 9 October 2007 in Macao. Thelma Kay, Director of the Emerging Social Issues Division of the United Nations Economic and Social Commission for Asia and the Pacific (UNESCAP), declared in an opening speech:

> for the first time [at the Second World Assembly on Ageing], governments agreed on the need to link ageing to other frameworks for social and economic development and human rights, recognising that ageing will be the dominant and most visible demographic phenomenon in the present century.
>
> (Thelma Kay, UNESCAP)

Besides examining the region's changing demographic dynamics and progress made in implementing the Madrid Plan of Action, the high-level meeting identified priorities for further actions in three key areas: older persons and development, advancing health and wellbeing into old age, and ensuring enabling and supportive environments. In Asia and the Pacific, the number of older persons is growing rapidly, from 410 million in 2007 to about 733 million in 2025 and to an expected 1.3 billion in 2050. In terms of percentages, older persons will constitute about 15 per cent of the total population in 2025 and up to nearly 25 per cent by 2050, from over 10 per cent now.

Population ageing brings significant economic and social challenges, for developed and developing countries alike . . . Sobering statistics show that some 80 per cent of the world's population are not covered by social protection in old age. Finding ways to provide economic support for a growing number of older persons, through sustainable pension programmes and new social protection measures, is a daunting task, particularly in developing countries.

(United Nations Secretary-General, Mr. Ban Ki-Moon, in a message marking the International Day of Older Persons, 1 October 2007, http://www.unescap.org/esid/psis/meetings/AgeingMipaa2007/index.asp)

tuberculosis, other respiratory infections, malaria and so on decline with development, but chronic degenerative diseases increase. Philips and Verhasselt (1994) highlight this trend for the case of Thailand during the 1980s based on verifiable health data. Pneumonia, tuberculosis and diarrhoeal diseases sharply decreased, whilst diseases of the heart, cancers, accidents and poisonings increased.

However, the demographic or epidemiological transitions are merely models against which we must assess a rapidly changing balance of health and development for each location, and this is very varied. Recent data from the World Health Organization (WHO 2007) show that the transition from declining infectious disease to increasing degenerative disease has got stuck in many developing countries. In those instances, infectious diseases that in earlier decades were considered likely to be gone from the world by the twenty-first century continue to affect millions each year. For example, it is estimated that over one billion people each year still suffer from debilitating diarrhoeal disease. Meanwhile globalisation and urbanisation of environment, society and the economy have led to an increase in chronic health complaints, such as cancers, diabetes and obesity. Obesity has been declared one of the UK's greatest threats to health, putting extensive pressure on the health service and resulting in a Government White Paper on the topic in 2005. At the same time, degenerative diseases have increased in many developing areas as well as the high income countries. The consequence is that developing countries are currently experiencing the impact of both the diseases of poverty and the diseases of overdevelopment at the same time (WHO 1997 and each year onwards).

Beyond human infections lies a further category of livestock and crop health threats that in themselves can bring on emergencies regionally and globally. Animal diseases that have presented significant threats of a pandemic disaster in recent times have included foot and mouth disease, bovine spongiform encephalopathy (BSE), severe acute respiratory syndrome (SARS), and avian influenza. Whilst the first of these can decimate entire livestock economies, variants of the other three, namely in the form of Creutzfeldt-Jakob Disease (CJD), influenza H5N1 and SARS, are direct threats to human health. The last two of these can be transmitted person to person through respiration, and influenza was responsible for one of the greatest die-outs of the human race ever recorded in 1918. The estimates of deaths from this were between 20 million and 40 million people. Many of these died from secondary infections such as pneumonia. Although we could treat that these days, estimates for the UK alone should there be another serious outbreak are in the order of 750,000 deaths. Other livestock related diseases that have disproportionately impacted on the developing world include trypanosomiasis, which led to the displacement of millions of people escaping the setse fly that transmits sleeping sickness from cattle to people (Richards 1985). Chicken cholera, otherwise known as Newcastle disease, currently decimates a fundamental part of the poultry diet of millions of people in some of the most food insecure parts of the developing world.

Crop diseases have also had widespread impacts on food and nutrition security, such that where crop diversification is not maintained people become more vulnerable to total loss of livelihood and the means of survival. Genetically modified crops provide resistance to some plant diseases, but are often not sustainable in the longer run if they demand more nutrients from the soil or that growers become dependent on seed from businesses that control patents and commodities. Furthermore, the use of pesticides has increased exponentially in agricultural practices around the world in a bid to increase output by controlling pests and diseases. Pesticides are similarly a marketable commodity that can be sold unscrupulously. Local growers therefore often become locked into a system of using more and more fertiliser and pesticide to maintain their comparative advantage as producers, whilst becoming dependent on suppliers. Combined with a lack of knowledge about these as hazardous substances, human health becomes compromised.

This introduction therefore suggests that health and development issues become disasters in relation to derivatives of environmental, social and economic factors. Some of the varied foci of this field in relation to sustainable development can be viewed in Table 4.1.

The relationship between health, development and disaster is a vast field of study and readers are also directed to work by Barrett (2004) and to

Table 4.1 ***A health and sustainable development perspective of disaster reduction***

	Underlying disaster and development themes		
	Ecosystems	*People*	*Economies*
Purpose	Ecological viability, sustainable human and ecosystem wellbeing and continuity	Social efficiency, justice, sustainable wellbeing, aspirations of people, social resilience	Economic efficiency, sustainable production, robust solutions to basic economic needs
Policy rationale	Equilibrium, holism, evolutionary integrity, protect nature, ecological awareness and risk avoidance	Build community resilience, develop local institutions, empower people with knowledge, capacity and rights	Develop markets and internalise externalities, growth centred development and disaster reduction
Health component	Emergent disease ecologies and health hazards, epidemiology and health of people and planet in contexts of environmental change	Human vulnerability to disease and ill health, local health management and knowledge systems, health behaviour and perception	Institutional development and global health, health care, environmental health, and emergency health care infrastructure and policy

Source: Adapted from Collins (2002:265).

Brown *et al.* (2005) on sustainability and health. Across a wide range of discourses on health there is a distinction between preventative and curative perspectives. The balance between the two is as old as conceptions of health itself, since people realised the association between poor environments, contamination and prevention. Curative medicine, though an essential part of development, has the limited, though essential, role of making us well again once a bout of ill health has occurred. Preventative and curative approaches can, however, be part of the same system, and as such are rarely reducible to a strictly medicinal formula, requiring changes in economies, the environment and society.

Other than during the emergency response moments, health and disaster studies are surprisingly sparse. There is a suggestion that there has been a lack of attention to the longer-term impacts of disasters on health (Cook *et al*. 2008), though thc work by Noji (1997) would be an exception, and this comment does not extend to the case of many health personnel who find themselves in disaster situations. Conversely, the extent to which better health helps reduce the impact of a disaster has been less examined. However, the resilience approach to disaster reduction provides a logic within which this relationship can gain more credence. In terms of examining health practice in disaster management, it is on the one hand about reducing health risks in the prevention mode, whilst on the other an integral part of emergency response systems. Health as part of disaster response is well covered in the various manuals of humanitarian relief and development. Two that provide a good overview are *The Oxfam Handbook of Development and Relief* (1995 and later editions) and Médicins Sans Frontières (MSF) *Refugee Health: an approach to emergency situations* (1997). Guidelines for minimum standards on providing emergency relief, agreed in principle by a wider range of agencies, are contained in the Sphere Project publication on the *Humanitarian Charter and Minimum Standards in Disaster Response* (2000 and later editions). Beyond providing guidelines, this is more oriented to the evaluation of emergency relief. A further example is provided by Wisner and Adams (2003) for the World Health Organisation in *Environmental Health in Emergencies and Disasters: a practical guide.* For the case of mental health there was a noticeable absence of texts until a lengthy and widely consultative exercise produced the Inter-Agency Standing Committee *Guidelines on Mental Health and Psychosocial Support in Emergency Settings.* These are intended to 'enable humanitarian actors to plan, establish and coordinate a set of minimum multi-sectoral responses to protect and improve people's mental health and psychosocial wellbeing in the midst of an emergency' (IASC 2007: 1).

The balance of health risks

The balance of health risks involves hazards and vulnerability, mitigated by different actions under the broad group of activities that we refer to as 'health care' (Figure 4.1). If health care is thought of as capacity, we have an approach to health risks not dissimilar to that indicated in Chapter 1 for disasters more generally. The hazards are aspects of day-to-day life that can cause ill health, including infectious pathogens, toxic substances, climatic conditions, physical impacts to body, and psychologically, socially or economically stressful environments. Overall vulnerability to these hazards may be socio-economically imposed on people. Alternatively, they are selected through behaviour on account of mental wellbeing, education and personality. The other aspect of vulnerability may be on account of biological susceptibility related to environmental or genetic influences. Health hazards and vulnerabilities are mediated by the capacity of health care to counteract illness. Health care is in part a function of advances in science and technology, how society and political-economic circumstances provide and prioritise human health, and how we choose to care for ourselves and others. The choice aspect in this equation is often referred to as *human agency*. The economic, social and cultural constraints on people's choices can be referred to as the *structural* aspects. Some of the subfields that may be loosely identified with health hazards, vulnerability and care are also indicated in Figure 4.1. The weighting of influences between the health hazards, human susceptibility and health intervention strategies determining a health outcome vary from one disease or health condition to the next, from place to place and over time.

Endemic and epidemic infectious disease

Despite breakthroughs in understanding the biomedical and microbiological components of infectious diseases, the benefits cannot be realised without understanding the relative influence of human vulnerability to social, ecological and other health risks. By identifying interrelated influences on infectious disease, we are also better equipped to recognise, give early warning of and control the influence of proximate and underlying causes of ill health. Different aspects of infectious disease causation may then be prioritised for pathogenic, human or more contextual interventions targeting the right time, place and people. This approach to infectious disease reduction combines varied knowledge to identify points of intervention which produce potentially complex

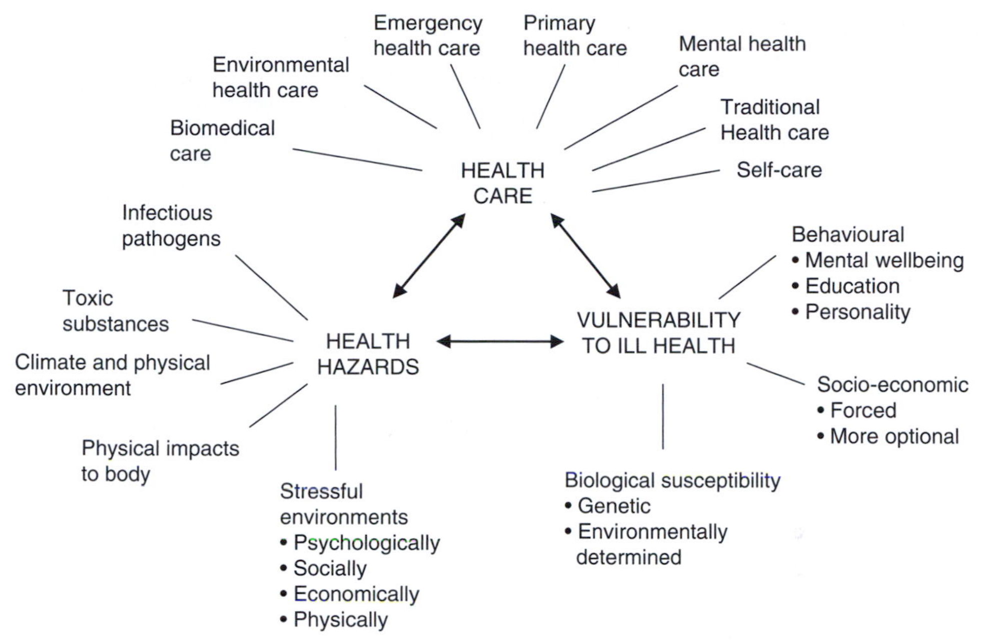

Figure 4.1 Health hazards, vulnerability and care

influences. It is an integrated view that also recognises the mixed interpretations of actual and perceived health risks. There is arguably a renewed demand to put infectious disease issues into an integrated disaster risk reduction framework with increasing recognition of its value (WCDR 2005). This is demanded because the magnitude of infectious disease in parts of the world is a vastly greater cause of morbidity and mortality than any other disaster type. The case of infectious disease in Africa presented in Box 4.2 puts the dimensions of health disasters into perspective.

Despite advances in primary health care and biomedicine in the global context, progress with vaccines and decreases in mortality from treatable illnesses, life expectancy remains distinctly lower in sub-Saharan Africa than most other parts of the world. Under-5 mortality is very high. Persistent incidence of 'old' diseases continues and HIV/AIDS constitutes a mass emergency with impacts on society and the economy. A scan of any World Health Organisation annual report of the last decade reveals that there have been significant improvements in controlling mortality from common infectious diseases. The issue is that mortality from common infectious diseases is still unacceptably high and that there remains a persistent and in some areas increased morbidity. This suggests

Box 4.2

Infectious disease in Africa

If we consider infectious disease in Africa as largely a development issue, then an immediate impression of the ratio between underdevelopment death and 'disaster' death might be estimated. Using data for 2003 from the annual reports of WHO (2005b) for health data and from IFRC (2004) for disasters related data, we can calculate that in Africa there were 757 under-5 illness related deaths for every one disaster death of any age. If we use the 15–60 years cohort for health data the ratio is even higher. However, as indicated in Chapter 1, disaster data are also largely about those affected (not just killed). In this instance, we find that for every one under-5 death from illness there were four to five disaster affected persons. In the case of the United Kingdom that year there were 164 under-5 deaths through a health complaint for every one any age disaster death. Data for some of the individual countries of Africa, the continent as a whole and the UK are included in Table 4.2. This also shows the contrasting amounts spent on health care in each of the countries.

Table 4.2 ***Health and disasters in Africa***

	*<5 Mortality rate per 1000**	*Adult mortality 15–60 yrs (male)**	*Total $ on health per capita**	*Total disaster deaths***	*Total disaster affected population***
Africa	40	?	?	5,810	20,951,399
Angola	260	584	38	40	831
Ethiopia	169	450	5	144	13,310,026
Mozambique	158	621	11	69	713,013
Sierra Leone	283	597	6	?	?
South Africa	66	642	206	135	2,527
Sudan	93	348	19	240	325,219
UK	6	103	2,031	2,045	?

Source: *WHO (2005b) and **IFRC (2004).

The causes of death amongst the under fives are shown in Table 4.3 in comparison to what are commonly rated as natural and technologically related disaster casualties. It is clear that most of the deaths are caused by infectious diseases. Note that of those in the neo-natal category up to 80 per cent are also known to be related to infectious disease.

Table 4.3 ***Causes of death for under five years of age in Africa, 2003***

	Deaths (n)	*%*
HIV/AIDS	285,000	6
Diarrhoeal diseases	701,000	16
Measles	227,000	5
Malaria	802,000	18
Acute respiratory diseases	924,000	21
Neonatal causes (inc. diarrhoeal)	1,148,000	26
Injuries, intended/unintended	72,000	2
Other	233,000	5
All deaths (WHO 2005b)	4,396,000*	100
Disasters natural and technological. (IFRC 2004)	5,810**	100

Source: *WHO (2005b) and **IFRC (2004).

that health programmes are getting better at treating symptoms and saving lives, but in terms of tackling the burden of sickness are still a long way off achieving adequate development of prevention. The purpose of flagging infectious disease prominently in this book relative to other categories of disaster is not to detract from the mainstream disaster reduction work based around 'natural and technological' disasters. It is, rather, to emphasise the importance of seeing health and infectious disease as a mainstream priority of the disaster and development agenda. Furthermore, beyond being a disaster category, health and disease can be associated with pretty well all the other disaster equations, for example from famine, food and nutrition insecurity, flooding and cyclones to conflict, industrial accidents and wildfires. Up to 80 per cent of diarrhoeal diseases can be associated with malnutrition in some regions (WHO 1997), floods intensify health risks (Few and Matthies 2006), conflict destroys health care, industrial accidents and wildfires spread pollutants, respiratory complaints or full poisoning. The latter is exactly what happened at Bhopal chemical plant, India, in 1984.

Infectious disease risk reduction

Infectious disease risk management (IDRM) seeks to reduce uncertainty in health disasters through integrated assessment and targeted intervention. It identifies cost effective ways of avoiding disease outbreaks waiting to happen, whilst reducing the impacts of those that ensue. This is because integrated assessments distinguish risks applicable to unique moments, locations and people, in respect of varying disease hazards and vulnerabilities. Moreover, reactions to emergent and resurgent diseases for contexts of 'intensive development' contrast with those of 'underdevelopment'. Emergent diseases are where there are new types of infection, such as those that are caused by a newly appeared or adapted pathogen. Resurgent diseases are those that have been present before, but which were dormant for a period before making a come back, either through having evolved a new strain, such as occurred with the 01 El Tor strain of cholera in 1961, or where human immunity has been weakened, as occurred with tuberculosis. Emergent and resurgent diseases often impact on society and national economies because of uncertainties about causes rather than numbers of cases. Thus, responses to BSE, CJD, Variant CJD (vCJD) and avian influenza contrast with those associated with mass incidence diarrhoea, tuberculosis, measles, malaria and HIV/AIDS. Different levels of 'acceptability' seem to accompany uncertainties about infectious disease risks. But integrated assessment of more certain diseases of underdevelopment may inform us about new health crises. Experience from economically poorer parts of the world suggests that IDRM can engage a wider community in addressing real and perceived risks. The assumption is also that investment in integrated disease risk reduction would assist transboundary preparedness and avoidance of future health disasters.

By reducing the uncertainty about infectious disease risks, epidemics are more likely to be avoided. However, the probabilities of some expected outbreaks such as avian flu are not known, demanding a whole new approach for decision making when risks are unknown. Infectious disease risk management may be considered the art of deciphering and negotiating between real and perceived health influences. It is also a science for establishing empirical evidence, knowledge systems and capacity for monitoring infectious disease risk indicators and responses. This approach can guide varied and effective intervention strategies that remain relevant to the extent they respond to the true nature of risks identified. Thus an environmental intervention may be appropriate for a disease in one context, vaccine in another context and community resilience building in another. Potential interventions for any one disease may be social, medical, economic, environmental, infrastructural, perceptual and often combinations of these backed up by education.

Infectious disease risk assessment therefore focuses on identifying reliable health risk indicators. Meanwhile, varied prevalence of infectious disease serves as an indicator of development issues, disaster preparedness and response capacity. As with disaster management more widely, the process of risk assessment and intervention is advanced through reflective learning in practice (see Chapter 5). In this respect there is vast experience already to draw on from a history of attempted removal or management of well-known diseases. Whilst uneven development is the clear underlying influence on infectious disease (Doyal 1987; Cairncross and Satterthwaite 1990; Leon *et al.* 2001), the relationship between uncertainty and 'acceptability' relative to established infectious disease 'norms' is less frequently acknowledged. Some diseases with unknown pandemic consequences impact on psyche, economies and society despite a limited number of cases. It would be inappropriate to advocate the scaling down of current preparedness and reaction to recent threats such as BSE/vCJD, SARS and avian influenza. An integrated assessment of newer threats can, however, be viewed and appropriately moderated in the context of the longer-term health disasters of our times.

The vulnerability of the poor is also defined in terms of being located in hazardous environments characterised by inadequate water and sanitation, poor air quality, social exploitation and deprivation of the means of protection against pathogens. The niche for pathogen adaptation, whether bacteria, viruses, protozoa, amoeba, parasites or so forth, is often infected people. Pathogens also survive within changing external environments, from which they are transmitted to people. The approach is influenced by an ecosystems model of disease balanced between pathogen, people (or host) and environment. Philips and Verhasselt (1994) use a version of this in conceptualising health changes through development.

This awareness of pathogen determined health, but in a context, has a good scholarly foundation going back many years. Epidemiologists, health geographers, environmental managers, planners and other disciplines have variously demonstrated awareness of the importance of understanding the disease hazard in context (May 1958, 1960; Prothero 1977, 1994; Stock 1986; Weiss and McMichael 2004; Connolly *et al.* 2004). How health risks are a function of both hazards and people's changing vulnerability to them can be clearly identified in the work of Bohle *et al.* (1994), Wisner *et al.* (2004), Collins (1998, 2003), and Bates *et al.* (2004a, 2004b). It also underpins models adapted by Colwell (2002) with a more microbiological emphasis, and by Collins (2001, 2002) and Bates *et al.* (2004a, 2004b) drawing on traditions in medical, social and cultural geography, or aspects of epidemiology. The environment may be considered in these approaches as a physical entity such as a reservoir or transmitter of disease hazards. Additionally, it is a wider context of disease hazard interpretation. Multiple levels of interaction from pathogen to potentially the entire world system define an integrated approach for addressing infectious disease complexity. A basic representation of this larger view of the situation broadly emphasising hazards and vulnerability is presented in Figure 4.2. Pathogens have already been described earlier in this chapter.

Disease pathways are all of the routes by which a disease becomes transmitted; vectors such as mosquitoes, flies, fleas, rats can be joined by environmental features such as transport systems, water, the air, soil and so forth. These pathways or transmission routes link pathogens in environments with pathogens in people. Pathogens such as the *Plasmodium* species that causes malaria undergo change whilst in the human host as part of their life cycle in people and mosquitoes. There are many other examples, such as, for example, the cycle of the bacterium *Borrelia recurrentis* via lice or *Borrelia duttoni* via ticks, both of which

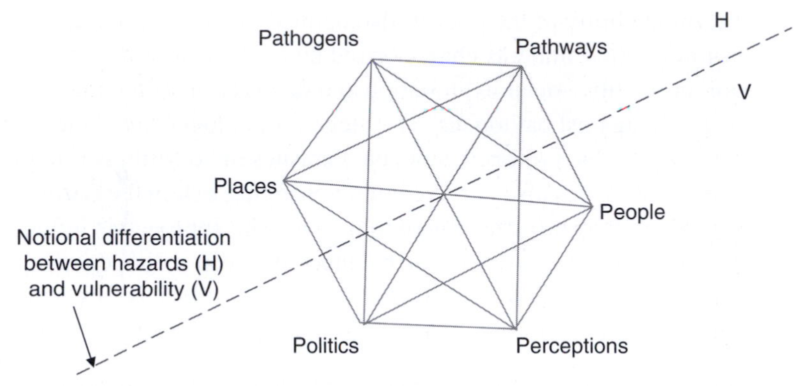

Figure 4.2 The health ecology approach to infectious disease risk reduction, for health security at global, community and individual levels

cause relapsing fevers. *Wuchereria bancrofti* are nematode worms transmitted from person to person via mosquitoes whilst in the form of microfilariae that undergo several changes. In humans, this eventually leads to the condition elephantiasis, or hydrocele, recognisable as extreme enlargement of body parts. Further examples are river blindness, or onchocerciasis, caused by *Onchocerca volvulus* and involving a female *Simulium* fly; yellow fever, involving mosquitoes and monkeys; trypanosomiasis, or sleeping sickness, from a protozoa, via flies and animals; schistosomiasis, involving snails; *Schistosoma*, water and cerariae; leishmaniasis, involving animals and sand flies; and plague, from *Yersinia pestis* bacteria via fleas and rodents. A full inventory is beyond the scope of this book. A highly accessible text covering the basic epidemiology of these and other disease cycles is contained in Nordberg (1999).

Some pathogens are transmitted directly to people from the physical environment, such as in the case of cholera, where the first stage of infection is through ingestion of *Vibrio cholerae* direct from water or when food has been contaminated. Once in the alimentary canal, the bacteria cause an extreme toxic reaction in the gut lining and life threatening diarrhoea. When it is not causing a disease epidemic through expulsion back into the environment in diarrhoea, it lies dormant within the aquatic ecosystem. There is a long and detailed record of cholera research demonstrating the complexity of how this pathogen adapts, survives and becomes virulent in people and the environment. (Colwell and Spira 1992; Islam *et al.* 1993; Drasar and Forrest 1996). It provides a prime example of the delicate balance between an environmentally based disease hazard and human vulnerability whereby displacement, conflicts or other disasters inevitably increase the risks it presents (Collins 1998, 2003). As environments change for the better with stability, good

Plate 4.1 Community volunteers cleaning the neighbourhood in Beira, Mozambique
Source: Author.

governance and a cared for environment, the threat of many infectious diseases is reduced. Hazard reduction at this level is therefore a function of dealing with pathogens, pathways and places.

It is possible to notionally categorise infectious disease influences much further by focusing more specifically on the vulnerability or resilience perspectives, but, as suggested in the model represented by Figure 4.2, it makes little sense to separate that entirely from pathogens, pathways and places as all are interlinked. With people included in the model, beyond merely bodies that are susceptible or immune to infection, we enter the realm of individual and group behaviour, choice, and mental and socio-economic wellbeing. Part of this is structurally governed, a consequence of politics. Poverty can be defined on this vulnerability side of the equation most readily in terms of people and politics. Reducing poverty makes people and society as a whole more resilient to the hazards posed by pathogens, disease pathways and changing places, such as, for example, in the context of changes in climate. Poverty, health threats and risky behaviour are relatively perceived phenomena. Different stakeholders in the health and wellbeing conundrum will present their versions of what constitutes a health influence differently depending on their perception of it and who they might be wishing to influence, and how. Hereby lies a risk of misrepresentation, misdirected funding and missed opportunity to reduce ill health.

Influences on health change on account of *disease ecologies* and *political ecologies of health*. Disease ecology was essentially the concern of pathogens in environments (Meade 1977; Learmonth 1988). Meanwhile, political ecology comprises political-economic and post-structural viewpoints inclusive of cultural, behavioural, environmental and developmental interpretations (Blaikie 1999; Peet and Watts 2004; Forsyth 2002), which can be applied to health. Awareness of multiple domains of infectious disease influences and interpretation has led to calls to extend disease ecology into political ecology (Mayer 1996), or to bring together 'multiple realities' through an ongoing review of unique health risks as part of an overall health ecology (Collins 2001, 2002). With this later example, an adaptive *health ecology* focus has been suggested dependent on different interpretations of moments of risk, different diseases and locations, and consequently varied intervention strategies. This translates in practice to ongoing monitoring of indicators to assess changes in health risks and building through experience the best means of targeting health interventions.

There are many implications for disease control practice through these approaches. For example, interventions may treat cases and control reservoirs (attack the pathogen), immunise or provide better nutrition (protect the host), or stimulate environmental improvement, personal hygiene or vector control (interrupt transmission). Eshuis and Manschot (1992) present a practical representation of this three-pronged set of options in the African Medical and Research Foundation manual on communicable diseases. But politics and the economy are a further part of the context, with varied perceptions and prioritisation of risk the drivers for proactive disease hazards and vulnerability management.
Some uncertainty in infectious disease causation usually has to be accepted as fundamental to these models. However, the problem of uncertainty in infectious disease control need not be an excuse for inaction. The next section considers the case of emergent diseases that have to date been associated more with middle to high income countries.

The case of BSE/vCJD, SARS and avian influenza H5N1

Bovine spongiform encephalopathy (BSE), from which Creutzfeldt-Jakob disease (CJD) and variant Creutzfeldt-Jakob disease (vCJD) appeared, first occurred in the UK in November 1986. By November 2002, 181, 376 cases of BSE had been confirmed. It is thought to have emerged due to a self-replicating protein (identified through a prevalence of abnormal prion protein), but it is possibly also a virus-like agent as it forms multiple strains. Uncertainty about BSE includes how BSE prions are spread and can contaminate. During foot and mouth disease (FMD) in the UK in the 1990s, mass destruction of infected cattle and sheep on economic grounds raised a scare that the burial of carcasses would pollute groundwater with BSE prions harboured by some livestock. A further concern was that prions might be released in smoke whilst carcasses were being burnt (HPA 2003). As such, two animal diseases together would cause greater impacts than the sum of their individual influences. The public pressure over the much larger FMD epidemic added to public perception of risks and tension concerning BSE. This escalated further, with a growing suspected link between BSE prions and human CJD. To prevent environmental contamination and possible prion transmission of BSE to people, FMD contaminated carcasses were not allowed to be burned if a cow was more than five years old. Environmental health officials were engaged in negotiating both a *theoretical risk* of transmission in burning carcasses and isolated evidence of linked BSE and CJD. The UK Department of Health (DOH) indicated a one in one million chance of contamination from burning 100 cows, with assumptions. The case demonstrates the power of uncertainty over policy in contexts of high economy.

Nonetheless, there has been about one case per million people worldwide of vCJD, with a total of 129 cases in the UK, so that it is not appropriate to belittle the importance of further work and precautionary principles. It is almost non-existent elsewhere in Europe, save for six cases in France, but there have been public inquiries and also new warnings. The Department of Health set up a CJD Incidents Panel in 2000. In July of 2005 it was confirmed that there was an increased risk of vCJD amongst those who had been operated on with instruments previously used on someone with vCJD or who received blood from people who were later found to be infected with vCJD (HPA-CDR 2005). Other risks have been variously indicated as potentially involving vaccines with bovine derived materials and animal derived pharmaceuticals, anaesthesia and tissue surgery. Some of this became apparent after monitoring CJD surgical incidents since August 2000, finding 183 practice incidents that *potentially* put people at risk. As a result fifty-seven patients were thought to have been potentially exposed to CJD (HPA-CDR 2005). Investigation of vCJD looked for geographical concentration of cases (in Leicestershire), consumption of beef purchased from butchers where cross-contamination with BSE infected bovine brain might have occurred, dietary patterns and possible iatrogenic transmission. The uncertainty therefore was related to ecology, epidemiology and associated theoretical risks. The disease burden was greatest in terms of 'crisis' management' costs. Agriculture Minister Gummer tried to demonstrate on television that British beef was safe by publicly feeding his daughter a UK beef burger. However, this addressed little of the public uncertainty about the (new) disease. In summary, the BSE case demonstrates the high uncertainty, economic impact and reaction, in a context of disaster risk with low levels of cases.

The world was also on alert in 2002 with the appearance of severe acute respiratory syndrome (SARS). Reports indicate there were 8,096 cases and 774 deaths from SARS between 16 November 2002 and July 2003, with a recorded 5,327 of the cases and 349 of the deaths in China (WHO 2003; Liu 2003). The case fatality rate was reported at 9.6 per cent, 7 per cent in the case of China. Throughout much of this epidemic there was also uncertainty about its ecology and epidemiology. As cases rapidly started to appear in other parts of the world, respiratory transmission became better understood and emergency measures were taken to block it. The assumption was that this was a highly contagious pathogen that might spread to all continents rapidly should there not be concerted emergency action to stop it. The impact on travel and economies began to be significant, but the disease stimulated a huge international response and was brought under control. Uncertainty had a major impact in the early stages, with several countries remaining on high alert for a long period beyond the main epidemic. In this case an uncertain disease outcome was contained through a major reaction and huge investment of resources to control it.

The highly pathogenic avian influenza ('bird flu') H5N1 alert intensified in 2005. By April 2006 there had been 194 cases and 109 deaths worldwide (WHO 2006), almost exclusively in Asia, but with a few cases within the borders of Europe. Uncertainty lay in the propensity of the avian strain of flu to mutate when in close contact with human strains of influenza. H5N1 is thought to be capable of genetic reassortment with human influenza strains such as H8N4 either in people or in a suitable host, such as pigs (HPA 2005). However, there may alternatively be more risk of gradual evolution into a strain more conducive to human infection. Doubts as to the probability of either scenario lie at the heart of explaining the global reaction to bird flu. Flu epidemics spread fear when there is both uncertainty and lack of a supply of guaranteed effective drugs. Strong international reactions to the unknown were inevitable. The unacceptable risk here stemmed from avian flu potentially involving random genetic or evolutionary process (a changing disease hazard) combined with changing human susceptibility, whether through self-inflicted farm based risk taking or having a lack of drugs to cure it. Preparedness and response demand a level of integrated infectious disease risk assessment hitherto only narrowly applied. It demands evaluation learnt through existing pandemics of better-known diseases and current development contexts. Arguably, diseases should not be viewed in isolation, the issues that apply to one, such as poverty, environmental quality, attitudes and risk management, being in part relevant to others.

The case of cholera

In the case of cholera we have more than 150 years of experience to refer to. The disease causing agent, *Vibrio cholerae,* was first discovered in the middle of the nineteenth century some few years after the medium of transmission through water in London had been suggested by John Snow, a British physician who lived 1813–1859. The aquatic environment of

Plate 4.2 An open well at Beira, Mozambique
Source: Author.

Plate 4.3 An arsenic contaminated bore hole in Bangladesh. When a bore hole is contaminated, people are told to revert to surface water, which needs to be treated
Source: Author.

Vibrio cholerae was only subsequently confirmed, but, famously, when John Snow had the handle of a water well removed at Broad Street, Soho, in 1854, cases in that area halted. Public confidence was restored, even though exact explanations for the success had not been discovered at that time. What John Snow's episode in London in the nineteenth century demonstrates is that solutions to emergent disasters can be found without our understanding all influences or exactly how complex systems function. His correct action was to think laterally, maintain open-mindedness and draw on potential multiple explanations of disease risk, intervening whatever way possible. In this instance, the uncertainty provoked a correct response. Cholera has affected most parts of the world at some time or other, and is currently an ongoing hazard in countries with favourable conditions for either reservoirs of toxigenic *Vibrio cholerae* or its easy transmission. Some countries possess both of these conditions and epidemics persist. We now know most of the pathways of cholera transmission and survival (Collins *et al.* 2006), but the certainty of further epidemics has not prevented its toll of morbidity and mortality.

Cholera is both a disease with environmental and disaster associations and one of poverty and underdevelopment. Recently, studies carried out using infectious disease risk management (IDRM) frameworks sought to isolate groups of indicators of cholera risk spanning its ecology, socio-economic and behavioural influences. It is clear from this work, however, that it is not sufficient to rely on monitoring technology and strengthening formal health institutions without establishing community-level participation. Consequently, in the case of Mozambique, which in some years has had the highest level of incidence of cholera in the world, risk committees were initiated within the community so that the community's own risk assessment was encouraged. The implications of reduced uncertainty within the community, together with capacity and monitoring systems being in place, are that more targeted interventions for epidemic prevention can save expensive post-outbreak relief operations. However, a deeper issue is likely to be one of dealing with the entrenched 'acceptability' of these most preventable of disease risks. Whereas individual and institutional knowledge may be strengthened by scientific endeavours, acceptance of the capacity of people to combat the risks contributing to ill health can be a greater battle.

There is a lot to learn about disaster and development more widely from integrated health assessments. The medical, social, economic, interpretational and other contextual aspects of health can be readily approached through an integrated assessment. This, however, raises questions about the varied levels of risk tolerance for different health conditions in varying contexts. The distinguishing of 'development' from 'underdevelopment' is notoriously subjective in an increasingly globalised world. The conceptualising of some health conditions as *uncertain yet controlled*, in distinction to *certain yet uncontrolled*, implies vastly different standards of policy.

Climate change and health

Whether or not there is currently much evidence for the impact of human induced climate change on disease pathogens, recognition of the complex association between climate and health needs to favour pro-poor risk reduction approaches. Shifting patterns of influence of climate change on pathogens, vectors and environments are likely to remain a relatively minor health risk in comparison to changes in the vulnerability and

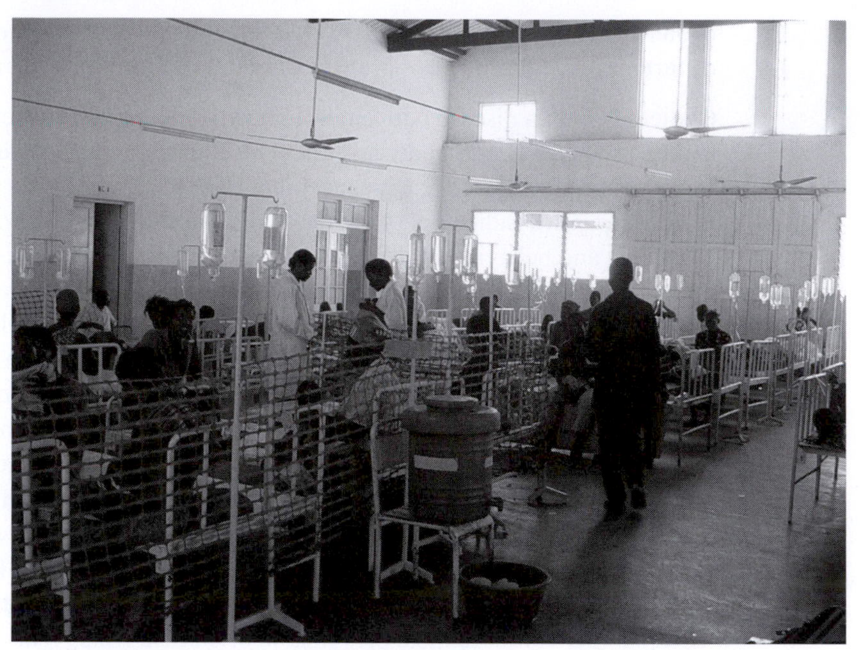

Plate 4.4 The cholera ward at Beira, Mozambique
Source: Author.

resilience of the poor. Meanwhile, some climate change influences could extend to impacts on mental health involving people across more varied development contexts. However, overall the discussion is essentially a political one, in that climate change and health raise issues about rights, representation and the development opportunities of those who would be most affected. A precautionary principle (Haremoes *et al.* 2002) consistent with the wider disaster risk reduction paradigm, and informed by appropriate health analysis frameworks, could be advocated for evaluating the complex association between climate change and health.

The links between infectious disease and climate change have been frequently flagged in the last few years, particularly with respect to cholera, malaria and the El Nino Southern Oscillation (ENSO) (Wilson 1995; Borroto 1998; Pascual *et al.* 2000; Bouma and Pascual 2001; Patz and Kovats 2002; Lipp *et al.* 2002; Colwell 2002; Pascual *et al.* 2002; Kovats *et al.* 2003; Lama *et al.* 2004). Sea surface temperature and the correlation with plankton blooms harbouring *Vibrio cholerae* are one such changing hazard, whilst global warming is considered to alter the latitudes at which malaria and dengue vectors can survive (IPCC 1995). Some further health associations are addressed by Haines and Patz (2004), and an overview is now encapsulated in the World Health Organisation statement on the Kyoto Protocol to the UN Framework Convention on Climate Change. The WHO has estimated that climate change has been responsible for 2.4 per cent of worldwide diarrhoea, and 6 per cent of malaria in the year 2000, and currently estimates that 150,000 deaths each year can be attributed to the effects of climate.

However, changes in health risk with different climates or rapid onset bad weather were explored in terms of seasonality well before current concerns were raised about the more unseasonable risks considered to result from human induced climate change. For example, the seasonal patterns of diarrhoeal diseases or malaria are a part of the recurrent reality of the poorest nations, such that local people begin to expect that they or a family member will be ill with these diseases at certain times of the year. Some respiratory diseases are similarly associated with a seasonal change in dry environmental conditions. Where the basic assets for life are available, people have adapted their own defences to changing seasonal risks, such as, for example, taking greater care over water quality during the wet season, or storing food during the productive period of the year to cover for a time of shortage in the dry season.

When changes in climate are unseasonable, people's adaptation to health risks can be off guard in terms of both awareness of the changing hazards and the material capacity to respond to new and changing health hazards. Nonetheless, in some more extreme instances of disaster it is important to remember that people have been found to be more resilient to health risks than normal. For example, during the great flood of Mozambique in 2000, thought to be associated in part with El Niño, fewer people than usual contracted major infectious diseases (Collins and Lucas 2002), only a fraction of the numbers that occurred during the economically successful and relatively disaster free year of 1998. Survivors of the flood of 2000 recounted how they had used great care in selecting their water, including rapidly thought-out cloth filtration systems for use on the roofs of the houses on which they were stranded. Also, post-Asian Tsunami, regions have reported only a fraction of the cases of infectious disease that were being predicted at the time of the disaster. One explanation again is the increased activation of defences against health risk around times of disaster.

Seasonal and unseasonal changes are related to a number of adjustments in the balance between health risks, involving changes in the disease hazard and human susceptibility. For example, fluctuations in diarrhoeal disease hazards and vulnerability are shown by Bardaj and Rao (1988) and Collins (1996) to be a function of changes in the seasonal cycle of rainfall, but controlled by variations in development, within and between both urban and rural areas. The wider role of seasonality in rural poverty was notably outlined by Chambers *et al.* (1981). Other studies examined changes in the exposure and vulnerability of refugees and displaced people in relation to the environment (Toole and Waldman 1993; Collins 1998). In each instance, the essential components are changing exposure to changing disease hazards (disease ecology) and changes in the resilience of the individual, communities and local institutions to protect against infectious disease. When hazards and vulnerabilities coincide, critical risk thresholds are frequently breached, epidemics occur and new endemic areas may be left in their wake. This is the story of the entry and persistence of many infectious diseases around the world, including the ancient scourge of diseases such as cholera, which shifted across Asia and since 1972 has also been endemic in Africa.

Changes in climate can clearly influence health by altering the distribution, magnitude and duration of spatial and temporal risks associated with exposure to disease causing agents. Meanwhile it also alters socio-economic vulnerability to ill health, particularly through changes in nutritional capital, and consequent biological susceptibility to infection. Furthermore, changes in psychosocial well being, which can also affect physical health and be affected by it, may be influenced by changes in climates. The physical risks are distributed unevenly between the rich and the poor, the poor being the least buffered against infectious disease hazards, such as diarrhoeal disease and malaria, and most dependent on the immediate productivity of local environments for a food supply. Richer people are less affected because they are less exposed to health hazards and have access to a more global marketplace. They would therefore have less immediate reason to be bothered about potential links between changing climates and health. The psycho-social impacts of climate change are more complicated in terms of ascribing their relative impacts on different wealth groups. But extreme climate events, such as flash floods, tornadoes and extreme droughts, impact on the poor *en masse*, whilst the rich are more exceptionally associated with this source of bereavement and longer-term trauma. Beyond the environmental, biomedical, socio-economic and psycho-social dimensions of the climate change and health nexus, the relationship is selectively represented in terms of different policy agendas. People and institutions use climate change to promote different political agendas. The relationship between climate change and health is therefore a complex one that we will struggle to fully account for in terms of its real, perceived or manipulated role over the coming years. A good review of ways into the climate change and health debate is also provided by Few (2007), who emphasises the need to consider the internal and external aspects of human vulnerability in this context.

Examples of a simple cause and effect explanation of health and climate change include the following: a rise in temperature equals more malaria, more diarrhoeal disease, more famine, and there will be more heat stroke. A shortcoming here is where a health problem in relation to climate change is taken in isolation from confounding variables, positive feedback loops from secondary qualitative variables and/or a lack of comprehensive data. For example, if the temperature rose in the Sahel and parts of Southern Africa, but due to global economic, security, governance and local development breakthroughs fluctuations in food and nutritional circumstances improved, then the potential impact on infectious disease (malaria, diarrhoea, tuberculosis) would probably be overridden by the improved socio-economic wellbeing. A further dilemma would be if climate change related thinking drew too much attention to an infectious disease hazard that we otherwise might adequately adapt to should non-environmental and non-pathogenic vulnerability be adequately reduced. Meanwhile, the lack of recognition (or acknowledgement) of the often less quantifiable psycho-social and/or political-ecological domains of health might similarly open doors to less effective intervention strategies, causing opportunities to address the underlying causes of health insecurity to be missed.

Understanding the changing distributions of health hazards and vulnerability with climate change may therefore be prone to overemphasis or underemphasis depending on different interpretations of a health risk. It is suggested here that acknowledging complexity is important in climate and health policy making, as this opens up the possibility of wider representations. The uncertainty of this relationship reinforces the need for application of a precautionary principle to climate change. This is simply to say that if we don't fully understand the nature of the risks, then we must be careful to ensure that varied interpretations are considered, including those 'scientifically' less likely. It is in this process that the real issues of vulnerability to climate fluctuations can be addressed, whether human induced or part of more regular environmental cycles. Furthermore, regardless of pollution's impact on climate, pollutants are in any event bad for your health. Improved environmental quality can be achieved through industrial management systems that respect health concerns, limiting the single agenda of overconsumption.

It has long been established that 'scientific truth' is often based on part-truths or whole truths evident from just some parts of a complex system. The democratic policy domain is therefore one that, although hopefully guided by good science, must ultimately also consider complexity in responding to the uncertain spaces that the interests of the majority occupy. This acceptance of the need to live with some level of uncertainty is, it might be argued, what lies at the heart of the wider disaster and development approach.

Global climate change is in any event clearly inevitable, and substantial change over short periods a distinct reality. Whilst rapid onset climatic events have occurred over the millennia, there may be an increase in the frequency of disruptive events and there are now billions more people around to be at risk. Furthermore, choosing to ignore the implications of changing climate and health associations in the more gradual sense could be tantamount to standing still whilst pathogens adapt, environments become more risky and people more vulnerable. However, choosing to engage with the components of this complexity now can potentially offer respite through the preparedness and effectiveness of health intervention strategies for years to come. Whilst the relationship between climate and health emerges as more complex than we yet realise, at times subject to exaggeration and at others inappropriately ignored, there is clearly a demand for the application of new health evaluation frameworks. Intelligent and dynamic approaches to health risk reduction, knowledge that empowers people to protect themselves, and knowledge for sustainable development and health are three of the keys to a precautionary agenda. This recognises complexity, is oriented to the needs of the poor and can guide effective health interventions.

An emphasis on human vulnerability and resilience rather than disease hazards has been more widely highlighted in integrated studies (Davis 1996; Kalipeni 2000; Collins 1998, 2003; Bates *et al.* 2004a). People are socio-economically vulnerable and consequently biologically susceptible when there is no livelihood security, where rights are denied, where there is conflict and when environmental disasters occur. At least two of these circumstances can be directly associated with climate changes, although they have not necessarily been proven to be human induced climate changes. Whilst infectious diseases invade the more vulnerable human spaces, people become susceptible through poor nutrition associated with poverty and dwindling food security. Notably, there is a well documented direct link between infectious disease and malnutrition (UNICEF 2003) and climate change, social vulnerability and food insecurity (Bohle and Downing *et al.* 1994). A health threatening cycle of poverty, loss of livelihood and nutrition security, and poor health can ensue when climate fluctuations cause crops to fail. However, changes in hazards and vulnerability, whether related to climate or not, happen as a by-product of unsustainable development, with implications for physical health beyond the infectious diseases. The implications for the chronic degenerative diseases in relation to climate change are currently less apparent, but there is clear evidence that colder winters kill more of the old and infirm, as can excessive heat waves such at that in Southern Europe during the summer of 2003. Climate fluctuation impacts on health tend to select the most vulnerable, whether through cold or heat, flood or drought, rapid onset or slow onset. Whilst the science of predicting when, where and who will be most effected is still in its infancy, the outcomes of climate changes once they occur are well understood.

Thc political issues are foremost, in that with anthropogenic inputs to climate change we must address the costs and benefits for health based on the combination of associations between climate and health, in an entrenched policy domain. The question is whether we are prepared to pay the price of some forms of development in terms of possible climatic impacts on health. However, this is an equation that cannot be quantified. Furthermore, it is right to continue to ask in this process, who is represented and who is not in making such decisions. There is clearly a moral, if not an economic, issue where the minority high consuming nations through their pollution potentially upset an already delicate balance between climate and health in the poorer nations. Whilst it is true that we do not know the impacts on the poor of climate change it is unjust to argue that people should adapt to change or cope. Coping after all for many already implies subsisting on the minimum for survival. Whilst the majority world is remarkably resilient and has learnt to cope with adverse climate, this cannot justify the continuation of potentially unsustainable development, the benefits of which are reaped elsewhere.

Furthermore, the politics of climate change dictates that the knowledge industry, received wisdom and 'informed comment' based on 'scientific truth' tend to also be based in those same rich enclaves of the world. The knowledge industry may switch sides on this issue frequently as new information and career opportunities emerge. Dependence on the 'figures' arguably creates a policy environment that can fall short of understanding the full impact of climate change, particularly amongst the world's disadvantaged nations. As it is the health of the poor that will be affected first, and in the greatest magnitude, even with conservative predictions of change, which political system should we look to as the guardian of these interests? Meanwhile, a precautionary approach is a simple concept that interprets well across different cultures and could be more ardently pursued in terms of reducing climate changing emissions.

Table 4.4 summarises how current IPCC (2007) projections for forthcoming climate change might impact on health. A closer analysis of processes of change in health risk and resilience using the integrated health security approach we identified in Figure 4.2 is provided in Table 4.5.

Table 4.4 ***Projections for extreme weather events for which there is an observed trend and their potential or likely impacts on health***

Climate phenomenon and direction of trend	*Likelihood of future climate trend based on projections for twenty-first century*	*Hypothesised impacts on health*
Warmer and fewer cold days and nights over most land areas.	Virtually certain	Increase in infectious disease incidence through spread and persistence of disease vectors and pathogens in areas where the cold previously prevented them.
Warmer and more frequent hot days and nights over most land areas.	Virtually certain	Increase in infectious disease incidence through spread and persistence of warm climate pathogens and vectors.
Warm spells/heat waves. Frequency increases over most land areas.	Very likely	Increase in heat stroke in temperate climates. Increase in infectious disease risks from hot climate disease vectors and pathogens.
Heavy precipitation events. Frequency (or proportion of total rainfall from heavy falls) increases over most areas.	Very likely	Increase in flash flooding and related breakdown in infrastructure, increasing health hazards, injuries, vulnerability and displacement.
Area affected by droughts increases.	Likely	Increase in food and nutrition insecurity. Climate related forced migration increases susceptibility and exposure to health hazards. Loss of livelihood assets increases socio-economic vulnerability to ill health.
Intense tropical cyclone activity increases.	Likely	Increase in rapid onset breakdown in infrastructure causes injuries and health hazards, vulnerability and displacement. Loss of livelihood assets increases socio-economic vulnerability to ill health.
Increased incidence of extremely high sea level (excludes tsunamis).	Likely	Widespread flooding increases health hazards, vulnerability and displacement. Loss of productive land through flooding and salination increases food insecurity. Loss of livelihood assets increases socio-economic vulnerability to ill health.

Source: First two columns are from IPCC (2007:7).

Table 4.5 ***Climate related risks to health based on integrated health security approach***

Health risk category	*Process of change in health risk and resilience in relation to climate*
Pathogens: i.e. bacteria, viruses, protozoa, parasites	Temperature and biogeochemical sensitivity
Pathways: i.e. vectors and environmental reservoirs	Distribution and viability of transmission routes of pathogens including via vectors (mosquitoes, flies, fleas, rats, snails, aquatic organisms, etc.) and environmental reservoirs (water, soil, phytoplankton and living spaces); hospitals (MRSA); food
People: i.e. susceptibility to infection relating to basic and extended needs	Temperature and water; nutrition security; psychosocial wellbeing. displacement; exposure to infection, including through drought and flood; socio-economic status and livelihood security; changes in availability of care
Perceptions: i.e. knowledge, attitudes and personality	Education; fear; experience; conscience; coping with uncertainty; media representations; changes in awareness and behaviour; changes in sense of community
Politics: i.e. political economy of health, advocacy and lobbying	Prioritisation of resources; politics of humanitarian aid, trade and environmental issues, including changing roles of international regimes, and conflict over natural resources; policies that prioritise preventative health actions and adaptation to climate change
Places: i.e. physical environmental quality, culture and local economies	Environmental quality through drought and flood related changes to water, land, air, vegetation; hazard modification in natural (i.e. land and water stability) and built environments (i.e. building, energy and water infrastructures); changes in economy and society

The case of HIV/AIDS

Acquired immunodeficiency syndrome (AIDS) has been the biggest world disaster of the last couple of decades. It is caused by the suppression of the immune system through infection by the human immunodeficiency virus (HIV). The origins of the pathogen responsible for it are unclear, but most likely for humans it arrived through transmission of the virus from primates. It was first noticed as a deadly disease in 1979 in the United States amongst homosexual men. Transmitted by the exchange of blood or other bodily fluids from person to person, it became identified with certain high risk groups such as the sexually promiscuous and drug users. However, it has also been associated with blood transfusions that were inadequately screened. As it has spread around the world it has increasingly been realised that it is a disease of which potentially anybody may be at some risk. This is because it can also be transmitted through conventional heterosexual activity and from mother to child. Protection is by way of sexual abstinence, use of a condom during sex and avoidance of any contact with blood that is contaminated with HIV. An account of the causes and consequences of HIV infection in a development context is succinctly provided by Whiteside and Sunter (2000).

The United Nations has a dedicated Joint Programme on HIV/AIDS (UNAIDS) that straddles most of the UN organisations. This approach further testifies to it being a multi-sectoral and interdisciplinary health issue. Some recent figures for AIDS cases from UNAIDS (2007) estimate the following global figures:

- The number of people living with HIV = 33.2 million
  - Adults = 30.8 million
  - Women = 15.4 million
  - Children under 15 years = 2.1 million
- People newly infected with HIV in 2007 = 2.5 million
  - Adults = 2.1 million
  - Children under 15 years = 420,000
- AIDS deaths in 2007 = 2.1 million
  - Adults = 1.7 million
  - Children under 15 years = 290,000

Every day, over 6,800 persons become infected with HIV and over 5,700 persons die from AIDS. It is therefore one of the most serious challenges to public health and one of greatest of human disasters the world has known. The story of AIDS has been one of blame, neglect and ignorance. The stigma caused by its association with high risk groups of people has never fully been overcome. For many parts of the world, the association with high risk groups, and representation of the disease in those contexts, only served to distract from the risks to society as a whole. The definition of a high risk group soon became regionalised in the mid - to late 1980s, such that the most economically disadvantaged parts of the world, namely sub-Saharan Africa, accounted for the greater numbers of AIDS cases. A more contextual account of why HIV/AIDS became a global issue and was regionalised in Southern Africa is addressed in some detail by Barnett and Whiteside (2002). Further regional data from UNAIDS/WHO (2007) substantiate the regional focus. They show that more than two out of three (68 per cent) adults and nearly 90 per cent of children infected with HIV live in this sub-Saharan Africa, and more than three in four (76 per cent) AIDS deaths in 2007 occurred there, also illustrating the unmet need for low cost antiretroviral drugs there (UNAIDS/WHO 2007).

The most significant legacy of HIV/AIDS has been its impact on societies and economies in the worst affected regions, causing delays to development and an orphan crisis (Guest 2003). In the process it has highlighted the plights of disadvantaged groups, gender inequality (Commonwealth Secretariat 2002), and misprioritisation and shortcomings in health preparedness globally. It has, however, also spurred one of the biggest funding responses to a health issue in history, with multiple foundations, international organisations and NGOs focused on its eradication, and prominent representation as a Millennium Development Goal. As the number of cases continues to increase in countries such as China and India, whilst only slight signs of improvement are yet evident in Africa, questions remain as to why the disease is still very much among us and what more might be done, beyond finance and information to reduce it. The response to this question would be the start of a whole new chapter beyond the scope of this text.

Food, nutrition security and famine

Food, nutrition security and famine are vast subject areas also requiring an entire volume. Here a short overview is provided to point readers in the right direction in terms of how to conceptualise the issues and identify this topic along the development to disaster nexus and back again. There are as many definitions of food security as there are explanations of what causes it. Thirty-two of the definitions are listed by Maxwell (2001). The multiple explanations for world hunger, with an emphasis on twelve of the myths, are addressed by Lappé *et al.* (1998). Meanwhile, Devereux (1993) has presented the multiple theories of famine. A multi-causal approach to nutritional issues is provided by Young and Jaspars (1995). Food insecurity can be very mild, barely detectable in the price regimes of food or perturbations in seasonal climates and shifts in the balance of livelihoods. At the extreme, we witness the case of severe malnutrition and famine. Between these two ends of mild food insecurity and famine lie a myriad of issues concerning availability, access, use, processing, choice and control of what people consume and are able to metabolise. Furthermore, whilst malnutrition leading to famine is a characteristic of economic collapse, climatic extremes and conflict, post-developmental and more economically prosperous societies are in the midst of an obesity epidemic. In some countries such as the US, UK and those of the Pacific Islands, and also in the urban areas of some low income countries, the obesity situation is considered of emergency proportions. Although lack of exercise and increasingly sedentary lifestyles are the lead cause of this epidemic, and for some there are genetic factors (particularly in the Pacific Islands), nutritionally poor diets and overeating have exacerbated the situation.

Three approaches already referred to in this book can be applied to food security, nutrition and famine:

- a multilevelled approach: from global food budgets to the nutritional status of the individual;
- systems analysis: the complex interaction of multiple influences on food and nutrition;
- hazard and vulnerability: shocks and trends that alter the risk of a food or nutrition crises.

The first has been used by UNICEF (1990) in its framework for the causes of malnutrition, in which the immediate or proximate causes of inadequate food intake and disease at the individual level are underlain by the household- and community-level access to food, social care, organisation and rights, public health and environment. These in turn are underlain by the basic causes, which concern political and economic infrastructure and resources. Looking at the different layers of food and nutrition insecurity provides an opportunity to consider the various levels of intervention. Interventions have been made in the fields of humanitarian relief, where a food insecurity crisis sets into motion relief that tries to quantify nutritional requirements and what supply of emergency aid can provide them. Moving from relief to development, the emphasis is less on nutritional balance and more on a healthy trading of produce. Where a reasonable range of food types is available, nutritional diversity is more or less taken for granted. Basic diets that are varied rarely produce nutritional problems.

Within the multilevelled approach in relation to access to food lies one of the influential comments on famine and food security associated with Amartya Sen, who in 1981 published his book *Poverty and Famines*. This emphasises the long held belief that a food crisis is not a question of availability, but rather a problem of access and interaction with food systems. According to Sen, this in turn is an issue of entitlements. Entitlements are production based (what people are able to produce), trade based (what they can buy), labour based (what they can earn) and inheritance based (what they inherit or are given) (Sen 1981). A series of macro-scale issues that in part is drawn from entitlement ideas is presented in Table 4.6.

Table 4.6 ***Food security: some underlying issues and options***

Issue	*Processes influencing food security*	*Intervention options*
Distribution of food and cash crops. Inequitable supply and access	Trade and changing terms of trade; globalisation	Trade negotiations, fair trade and debt adjustments
Fresh water supply	Environmental change, including climate change; conflict.	Environmental security through sustainable development and conservation; conflict resolution over water access.
Techno-agriculture	Large scale intensive farming; increased use of GM crops and pesticides	Avoid vulnerability of mono-cultivation; GM interventions based on as ethical, cultural, social or economic rationale
Land tenure	Changing rights of access to productive land	Land laws that respect the rights of local communities, empowering people to provide their own food security solutions
Ecosystem resilience	Ecological changes	Lower input sustainable agriculture; enhancement of indigenous methods; appropriate technology
Relationship of women to food production and management	'Man made famine', where links between food production and management and household nutritional security are jeopardised by constraints on women	Social, women's empowerment in marketing and budgeting for household food; labour saving strategies.

Examples of a systems approach to food security are numerous. For example, Box 2.2 (pp. 79–80) shows how food security is linked to livelihood security and how a web of factors link up to explain deficits in the two. Hubbard (1995) provides a detailed framework for a local food security system that links the food market, work and income, fuel and water with household decision making, and in respect of the nutritionally most vulnerable members of the household. He also includes the disease environment in this set of systemic processes. This is also a centrepiece of the UNICEF approach, which flags the close relationship between infectious diseases and nutritional status that we referred to earlier in this chapter. An inadequate dietary intake causes appetite loss, nutrient loss, malabsorption and an altered metabolism, leading to starvation. It causes weight loss, faltering growth, lowered immunity and mucosal damage, which is conducive to increased incidence, duration and severity of disease, which in turn causes even more appetite loss, nutrient loss, malabsorption and altered metabolism (Tomkins and Watkins 1989). The systemic aspects of malnutrition are all too prevalent throughout human life cycles. If there is a continued lack of an adequate food supply, underweight babies become stunted growth children, and the girls eventually become pregnant women carrying low rate of growth children.

Human nutritional requirements

That a core problem for nutrition security is not so much availability but access to food is proven by a simple look at some calorific calculations. Food energy is calculated in calories. On average, people need 2,100 calories per day to be healthy. Food availability for the world was estimated at 2,720 calories per person per day in 1990–1992 and has been rising ever since (UNDP 1997, based on data from FAO). This was up from 2,300 calories in 1961–1963. It has been projected that by 2010 even developing countries will achieve per capita food availability of 2,730 calories, but this is not the case in Africa, where daily per capita calorie intake has declined. However, food security is uneven within and among countries. Further to this, nutritional status reflects not only the quantity of food available and consumed, but also its quality.

Malnutrition is divided by the World Health Organisation into a number of categories, as follows:

- protein-energy malnutrition;
- iodine deficiency;
- vitamin A deficiency;
- anaemias;
- other nutritional disorders relating to lack of vitamins and micronutrients.

Each of these occurs in mass emergency situations, though the greatest in terms of rapid onset starvation is associated with protein-energy malnutrition. Hundreds of millions of people are estimated to lack access to food to meet their daily basic needs for energy and protein.
An examination of the data in various World Health Reports of the WHO would suggest that billions are thought to be deficient in essential micronutrients such as iodine, vitamin A and iron. The signs of protein-calorie malnutrition are kwashiorkor, which manifests itself as water retention in the legs, pale pigmentation, fat retention in the liver, head to chest ratio changes, and marasmus. The signs of Vitamin A deficiency are a brown spot on the eye that leads to night blindness, and eventually complete atrophy of the eye. Vitamin A is found in green leafy vegetables, orange or yellow fruit, butter and eggs. The sign of iron deficient anaemia is

lethargy and it can seriously affect pregnant women. Pellagra results from a lack of niacin, which causes skin diseases. Iodine deficient goitre is a swelling around the throat. For a full account of the problem of nutritional deficiencies, foodstuffs that contain what is required and recommended nutrient intake, further information can be found in the following:

- *The Oxfam Handbook of Development and Relief* (1995 and later editions);
- Médicins Sans Frontières (MSF) *Refugee Health: an approach to emergency situations* (1997);
- the Sphere Project publication on the *Humanitarian Charter and Minimum Standards in Disaster Response* (2000 and later editions).

Quantity and quality of food

Close assessment of nutritional status is a form of preventative action, as with information in advance some action can be taken to supplement people in a targeted way with appropriate types of food. In many nutritional crises food is available but not in the right variety of types. The energy content of carbohydrates, such as wheat, rice and maize, if taken on their own is very limited. People surviving on them alone must therefore consume large quantities. This gives rise to the spectacle of young children with swollen looking stomachs, but who are badly malnourished. If a food aid package contains some protein, such as beans and then oil added, then the quality of the food is immediately improved. One food that is often provided in emergency feeding centres to very malnourished children is UNIMIX. This is easy to absorb and can be used until strength is regained. An example of a recipe for this might include: 30 per cent maize (or corn) meal, 10 per cent oil, 10 per cent milk powder, 40 per cent beans (mashed or ground) and 10 per cent sugar. Famine early warning systems, risk assessment and nutritional measurement techniques are returned to in Chapter 6. Nutritional response and recovery procedures are returned to in Chapter 7.

Chemical poisoning

Chemical poisoning as a development concern has been highlighted since the early 1960s and the advent of the environmental movement. The slow onset of chemical poisoning in the world was the drive behind Rachel Carlson's book *Silent Spring* in 1962. Chemical risks to health are extensive, from respiratory diseases to allergies in areas of industrialisation, to the possibility of major incidents at chemical plants, spillage during transportation along roads and rail, or through sabotage. Chemicals are used in conflict, with dire results. Agent Orange was used in Vietnam to clear vegetation, but with the residue leading to a heightened chance of birth deformities there. Kellow (1999) provides a detailed account of the chemical risks associated with industry. Meanwhile, in agriculture the so-called 'Green Revolution' brought with it the intensified use of chemical fertilisers and pesticides in the developing world. The impacts on health of only some of these are understood, though there is increasing evidence that the amount of contamination that people are carrying has been increasing beyond recommended levels. Further concerns include the risk of disasters caused by acid deposition in rain, should this combine with other contaminants, airborne gases, particles and aerosols that cause climate change and radioactive pollutants. Some of these issues are addressed by Barrow (2005) in *Environmental Management and Development,* which is a further book of the Routledge 'Perspectives on Development' series.

Where atmospheric pollutants such as chlorofluorocarbons (CFCs) have destroyed the ozone layer, as in parts of the Southern Hemisphere, protection from ultraviolet rays from the sun has been lost and people exposed to daylight in those areas are subject to a higher than average incidence of skin cancers. Heavy metals such as lead in fuel and copper in and around smelting areas are a cause of poor health, particularly in children. Waste management is a vast field that in some countries such as the UK, has arguably become a crisis issue in terms of protecting the health of future generations. The problem would seem to be critical in terms of the lack of a solution from governments and an ongoing and worrying silence on this issue. Landfill sites are full and the many household and industrial items are still not being recycled. The problem of recycling extends to nuclear waste, which will remain a threat for hundreds of years to come. It is at this point that it is reasonable to consider the discipline of environmental management and disaster risk reduction as overlapping significantly. Common to both is a crucial role for the identification of a mechanism to obtain environmental justice for those forced to tolerate health risks imposed by others.

Mental wellbeing, trauma and coping

There are a number of different emphases in the broad field of mental health and disasters. Mental illness and psychological trauma are balanced by coping as an antidote. This line of study is often referred to as psycho-social, because it is not possible to separate mental and social wellbeing. There is generally agreement that disaster impacts on mental wellbeing, may cause trauma and that people experiencing disaster are best thought of as survivors rather than victims. There have been attempts to model what might happen to people mentally during and after disasters. Examples are provided by Hodgkinson and Stewart (1998), who show links between intrusive memories, cognitive appraisal and new dysfunctional beliefs to explain post-traumatic stress reactions. Beck *et al.* (1985) conceptualised the cognitive component of anxiety that Tapsell and Tunstall (2006) applied to the mental health aspects of flooding in England and Wales. They indicate a number of factors that can result from a flood disaster including the following:

- interpersonal relationships and family strains;
- financial stress and obligations to provide financial and social support to others;
- occupational stress;
- isolation;
- impacts upon personal identity and a sense of 'self' and 'place';
- attachment to home and possessions;
- loss of security in the home.

(Tapsell and Tunstall 2006: 103)

Dolce and Ricciardi (2007) provide evidence that there is also a psychological impact on disaster rescue volunteers, but in particular those who have had inadequate training and experience. The topic is, however, beginning to be addressed in a little more depth by some agencies.

Western biomedical concepts describe trauma as an individual experience (Pearlman and Saakvitne 1995). However, Summerfield (2000) and others (Summerfield and Toser 1991; Summerfield and Hume 1993) argue that medical models are limited because they do not embody a socialised view of mental health. Exposure to a massive disaster and its aftermath are not generally personal or individual experiences. It is in a specific social, political, cultural, economic and environmental setting that those who need help reveal themselves and the processes that determine how victims become survivors are played out over time (Jayawickrama 2005, personal communication). The psychosocial links between a disaster event and health are far from simple. Following extreme climate related events such as a flash flood, hurricane, fire or other forms of disaster, we know that people may be left traumatised by the loss of life and of livelihood (Desjarlais *et al.* 1995; Hodgkinson and Stewart 1998; Pupavac 2001), but the means of coping might be highly varied. Trauma response units are available in developed nations to deal with rapid onset disasters. Problems arise when the approach used in one part of the world is exported to another as the culture and meaning of mental wellbeing vary.

One under-researched area that should also be mentioned here is the direct links between physical and mental health. It has long been felt from both biomedical and socio-behavioural points of view that one effects the other. Many people acknowledge this without necessarily identifying the exact physical processes involved, though a combination of changes in the central nervous system, endocrine system and behaviour is the most likely. Slow onset changes in the environment leading to stress may have a similar effect when people's lives become displaced, such as through drought induced crop failure and forced migrations from extreme weather events. There is a growing awareness in the wider field of psychosocial care that the responses need to be increasingly about addressing community resilience through livelihood and human security rather than the traditional approach to counselling. This is not least because the latter would be ineffective in the face of mass traumatic events including too many people to council individually.

It is also worth mentioning here that even significant changes in air temperature can affect people for better or worse in a psychosocial sense. A lack of sunshine is a recognised cause of depression in northern countries, diagnosed as seasonally affected disorder (SAD). Some predictions on shifting climates point to an increase in cold grey weather in some northern latitudes should the North Atlantic gulf stream be headed off by southward moving cold currents from the melting northern ice cap. Any of these links in turn could have effects through behavioural changes, including people's sense of vulnerability, whether the hazards and risks are real or perceived. As with physical health issues stemming from climate change, there may also be reason to identify the psychosocial implications in some regions in terms of their impact on demand for health care services and related costs. Whilst the caring professions in the rich corners of the world might be partly equipped to address psychosocial issues, little or no provision is available in the majority of the world.

Trauma is thought to cause parts of societies to be less productive and less able to deal with further stresses. However, it is increasingly observed also that survivors of disasters and the displaced might not be negatively affected, but rather become strong and resourceful. The secrets to this are not well understood, but appear to be linked to resumption of the normal rhythms of life and locally sensitive caring. Whilst necessity is the mother of invention, extreme events can nonetheless make people dysfunctional mentally and physically.

Health care, social care and self-caring

The good news is the success with which humankind has tended to come up with a number of effective responses to health risks. Although they have not provided encompassing solutions to many ongoing health crises, primary health care, immunisation, environmental health care, food supplements, prevention, preparedness and control have advanced significantly. However, as was suggested in Figure 4.1, there is a wide range of health care options available that can mitigate disaster impacts on development, and development induced disasters. As health risks and outcomes change from place to place and over time, what approach can be developed for sustainable health care practices? To answer this we need to go back to the WHO definition of health that we started out with, i.e. 'a complete state of physical social and mental wellbeing'. If this and our recognition of complexity in health are taken on board, then health care by default needs to be very diverse in approach.

Some clear guidance can, however, be taken from the past. A key shortcoming has been and remains the vastly uneven development of health care between and within different regions. Jones and Moon (1996) highlight the 'inverse care law', in which expenditure on health care is highest where the need is least and lowest where the need is greatest. This is borne out by repeated World Health data each year showing that there are at least 8 times as many active nurses, 20 times as many dentists, and 30 times as many physicians per 100,000 people in Europe than for the same number of people in Africa. Comparison of drug availability can produce some equally startling statistics.

The legacy of ineffective health care development was also reflected in the colonial systems that tended to be oriented first and foremost at curative medicine for expatriate communities throughout the colonies. In many cases a system of apartheid in health care meant that more specialist services were only available for the European communities, with completely separate hospitals for indigenous populations. Both the need to concentrate specialist units where there were more European populations and the tendency for doctors to live in urban areas meant that many rural health care services in developing areas were left to residual parts of health budgets, missions and traditional health care practices.

Curative approaches to health are often associated with centralised health care approaches, partly similar in terms of governance to centralised responsive modes of disaster management. As with disaster risk reduction, what is needed in the health response for the masses of people beyond the formalised centralised system is more locally sensitive and widely accessible approaches. This was recognised early on in the primary health care approach emphasising equity, self-reliance and prevention. The eventual mainstreaming of primary health care into global health began to, in principle at least, address disparities in health within and between nations and other parts of the world using 'health for all' as its mantra (WHO 1978). The following definition clarifies what was meant:

> essential health care made universally accessible to individuals and families in the community by means acceptable to them, through their full participation and at a cost that the community and country can afford. It forms an integral part of the country's health system, of which it is the nucleus, and of the overall social and economic development in the community.
>
> (WHO 1978: 34)

Significant advances were made in developing primary health care strategies, with selective primary health care used to prioritise the bigger issues of the moment. Underlying the approach has been the belief that if low cost ('Western') health care could be adapted and made appropriate and accessible to the majority of the world, then ill health and disease would be brought under control. This would be the driving force behind an epidemiological transition. The Declaration of Alma Ata (1978) stated that primary health care should include:

1. education about prevailing health problems and methods of preventing and controlling them;
2. promotion of food supply and proper nutrition;
3. an adequate supply of safe water and basic sanitation;
4. maternal and child health, including family planning;
5. immunisation against infectious diseases;
6. prevention and control of endemic diseases;
7. appropriate treatment of common diseases and injuries;
8. provision of essential drugs.

Despite immediate and sometimes ongoing progress evident on these fronts of the PHC agenda (Lafond 1995), it has not yet been sufficient to overcome the impact of underlying development problems and disaster impacts on health. One explanation may be that despite much good work in PHC, it remains part of a global strategy (Larkin 1998) that cannot be sufficiently sensitive to the varied nature of changing health risks. There are internal contradictions between national health services and community participation in health. Curto de Casas (1994: 240) referring to case studies in South America, compared with local health services and found that, whereas national health services used curative medicine based on a physician-patient relationship, local health services used preventative medicine based on a community-environment one. Participation was possible under the national health care services, but was a complementary and controlled community participation, rather than community participation in execution and evaluation. The priorities were given by professionals, rather than being considered by the community.

There has also been a call for new partnerships between the various institutions attempting to promote or adapt PHC. The approach made an impact, but with health issues, particularly infectious diseases, in several parts of the world remaining out of control and life expectancy going down in some areas, there is frequently a call for new innovations in health interventions. The Bill and Melinda Gates Foundation has committed itself to looking for alternatives on the basis that 'grand challenges' in health require greater risk taking and 'creative, unorthodox ideas'. Overall it is maintaining of people in a poverty trap that underlies much of the social, economic, environmental, physiological and pathogenic determinants of ill health, demanding ongoing reassessment of what is exactly meant by PHC. There is a concern that new approaches are now critical if Millennium Development Goals are even to be partly achieved. Health is represented specifically in three of the eight goals, and makes a contribution to the achievement of the others (WHO 2005c). A full review of the situation in 2006 and some reasoned projections are provided by Dodd and Cassels (2006). Meanwhile, Whitehead and Bird have drawn attention to one of the significant problems as being lack of case or inappropriate care and the need to tackle the 'medical poverty trap' (2006: 396).

People centred health care

Development paradigms have made a fundamental change to the way health interventions are understood and promoted. Notions of development from within and giving a voice to the people were not new concepts in many cultures, but were often neglected by policy makers (Chambers 1997; Blackburn and Holland 1998). In recent decades 'participation' has represented development discourse and to some extent practice with a strong paradigmatic front. There is a consequent influence on approaches to public health policy such as that made evident by Cornwall *et al.* (2000). The burgeoning of the non-governmental sector health and development work already trying to promote 'community involvement in health' (WHO 1991) has provided a basis for participation. In a practical sense it led to the development of the methodology of participatory appraisal, which is now fundamental to household and community health interventions. It is, however, widely realised in health and development work that notions of 'participation' have meant different things to different people and styles of programming.

Finding out what people think, need and want, and therefore what might be effectively implemented using participatory appraisal, has only been a first step towards creating a sense of ownership of the process of health interventions. The further step is recognition that what people already do and think can be promoted as good for health using their own resources, or modified to that effect. There is added benefit in this approach as those at risk are potentially the same people as those implementing the strategy to remove the risk. This approach to participation is more empowering, sustainable and effective than imposing intervention strategies that are unfamiliar to the user. It is where the intended beneficiaries of a health strategy are recognised as having a voice and a right to own their own method of control. Community ownership is, however, an approach which is surprisingly less evident in many of the health and disaster reduction programmes. Also, the more precise manner in which the government and non-governmental structures can help facilitate sustainable health solutions in this way is often not well promoted. This is because the science surrounding health hazards and the sociology and politics surrounding community development have unfortunately tended to remain separate fields. Beyond combining disciplinary awareness lies the even greater task of shifting investment from a reactive and 'curative' response driven agenda in the health sector to one of risk reduction based on listening to people's ideas about risks. The uncertainty in the health disaster risk reduction field challenges health and development methodologies and techniques. It is likely that accepting complexity and uncertainty, and also embracing participatory and empowering approaches, is not expedient for all of the gatekeepers with command and control of the health service.

Health promotion

Health promotion is broadly a strategy for purposively promoting the health of a whole population, enabling people to increase control over their health. Once again there are parallels with those versions of disaster prevention and response that advocate local, community and individual preparedness. It is important to distinguish those aspects of health promotion that are more external to the individual, the structural aspects of health care, from those which stem more from within the individual. Whilst not mutually exclusive, the outcomes of health promotion are on the one hand more structurally determined and on the other more behavioural. Health education is about having the right knowledge, which is accessible to people and tailored to different traditions of health response. Policy on exactly what and how much to promote is determined by factors such as GNP and stability of the nation-state, political ideology, cost-benefit analysis and risk assessment, availability of trained personnel, organisations and information flows, and support from powerful groups such as medical professionals.

The less structural health belief approach contrasts with this significantly by emphasising the importance of behaviour, culture and psychology. It is an approach that has been referred to as a health belief model, and versions of this are quite well established amongst public health bodies, more formally at least since the 1970s. The model suggests that a person's likelihood of engaging in a certain behaviour is a function of their perception of relationships between actions and illness, their own susceptibility, the seriousness of the illness, and the costs and benefits involved. It is argued that understanding this is also crucial to successful (i.e. sustainable) health intervention and improvements. A more detailed outline of the structure of the model has been provided by Buchanan (2000), who included the overall themes in the left-hand column of Table 4.7. A series of influences (independent variables) important in promoting health belief in terms of these main categories have been added.

Self-efficacy in this context may be considered what people can do for themselves to mitigate adverse health shocks, whilst self-care is more about how people self-manage a sustainable response to their own health needs. This is very important to understanding the roles of 'coping' and of 'health resilience'. As health budgets fall far short of meeting demand, questions are raised about the role and capacity of the nation-state to address health needs. However, a further concern is that 'self-care' and some participatory approaches or endogenous approaches can in effect offer governments an opportunity to retreat with an escape clause from their responsibilities to address state care infrastructure and rights to support.

Ultimately, health intervention fronts are informed not only by what people are able to do to prevent ill health, but what people believe to be the cause of their illness. Health researchers working in the context of

Table 4.7 ***Influences on person-specific health outcomes***

Real or perceived characteristics relating to health condition	*Key influences*
Susceptibility	age; gender; money; class; nutrition; social status; environment (social, location, place); education; behaviour; genetics; immunity-access; entitlement; rights; beliefs-tradition
Severity	area; extent; morbidity; mortality; rate of incidence; frequency; levels of importance bestowed by politicians and/or health professionals
Barriers to changing behaviour	lack of co-operation; attitudes; values; level of comprehension; ability to plan ahead (and willingness); language and communication; lack of choice; entitlements; rights; vulnerability and sense of vulnerability
Benefits to changing behaviour	healthier life; longer life; increased stability for work; not transmit to others; improved mental wellbeing; empowering/status building; breaking the health/poverty cycle
Cues to action	type of health care system; advocated by community; type of media; community involvement in health; educational infrastructure; government stakeholder policy (all levels)
Self-efficacy	accountability; level of communication; future plans; level/propensity to awareness; conscience; self control; notion of responsibility and respect

Source: Main categories in left column are from Buchanan (2002).

implementing health improvement programmes, have made some contribution to exploring health perception issues. For example, Winch *et al.* (1994) have looked at the seasonal perception of malaria and Smith and Watkins (2004) at the perceptions of risk and strategies for addressing HIV/AIDS in rural Malawi. Risk perception is in fact an entire field in its own right, as demonstrated by Slovic (2000). However, the contribution of risk behaviour research to improving health interventions in the majority world still has a long way to go to make a real difference.

The relationship between people's knowledge and how they respond to what they know in terms of protecting against ill health was alluded to in the previous section. More specifically, water, sanitation and hygiene programmes, HIV/AIDS prevention projects, TB treatment programmes and malaria control programmes sometimes carry out a knowledge, attitude and practice (KAP) survey, or aspects of this, as part of the project planning monitoring and evaluation cycle (this is returned to in Chapter 5). This can be ethnographically based, although some wider quantitative representations are also useful. It is possible to survey health knowledge in a given area and group of people and then gain a reasonable impression about changing practices with changing knowledge. However, the attitudes part of this equation can be difficult to assess. The KAP assessment is dependent not so much on knowledge *per se*, but on perceptions of the issues based on people's pasts, and their attitudes to risk taking influenced by a number of factors. A perception of a risk, such as the high chance of becoming HIV positive through unprotected sex, might change behaviour to some extent, but there is evidence that, even with knowledge, risk taking continues. This is in fact one of the biggest challenges facing the campaign against HIV/AIDS. Whilst poverty can partly explain the high incidence of HIV/AIDS, there is growing comment from African scholars and practitioners that the risks have more personalised, cultural and psychosocial explanations extending beyond poverty and political economy. Some parallels can be drawn with people who continue to smoke knowing that there are significantly heightened risks of lung cancer. The subject matter leads to psychology and medical sociology and the possibility of being able to predict behaviour. Only limited work would appear to have addressed the applied alternatives to the health belief model (Montano 1986).

A condition of fatalism, perhaps due to previous disasters, can cause an individual to reduce their overall struggle for health and increase risk taking. It may of course be possible to trace structural factors in people's backgrounds here, but the more immediate influences occupy a behavioural space of attitudes, cultures and various addictions. What this further reinforces in the context of this chapter is that a people centred and more localised approach is the minimum requirement to be able to address attitudes. In many instances change will be dependent on individual mental wellbeing and decision making. However, the extent to which 'self-care' is something that people choose, adapt to, or have forced upon them by necessity remains to be fully researched (Edgeworth and Collins 2006). It is a question that is grounded in the bigger question of who is really responsible for health, development and disaster risk reduction.

Primary and emergency health care systems

We might further note the significance of the primary health care sector for disaster risk reduction by noting that primary health care principles include *equity* and *self-reliance* as well as *prevention* (WHO 1978). This includes 'at risk' people identifying the influences on ill health themselves (i.e. human agency) and being empowered, where possible, to adopt a self-care or community care approach rather than relying on more centralised disaster response systems (i.e. structural responses). This should constitute an intervention process prioritising in terms of the nature and context of ill health and interventions targeted at the right

Plate 4.5 Poster to feed back information on health risks to a community in Bangladesh
Source: Author.

Plate 4.6 Sharing a risk analysis with part of a community at Beira, Mozambique
Source: Author.

moment, in the right place and at the right people. The range of interventions available is vast, including environmental health care solutions (i.e. water and sanitation), education, diet and nutritional strategies, security, land tenure, management and governance reforms, immunisation, medicines, health care centres and workers, working with traditional health care and so forth. These responses can operate at any stage of the disaster management cycle, as prevention, mitigation, preparedness, relief and recovery. In the context of relief and emergency response it is also necessary to have people on hand who are trained in mass casualty preparedness.

One of the key questions in looking at health from a disaster and development perspective is the extent to which a health care issue is an 'environmental' and/or 'social' care issue. For instance, health and environmental care are closely associated both in theory and in practice. Health care systems provide useful metaphors and real examples for environmental management. In disaster situations, health care and environmental health care often become part of the same response initiative. Prioritisation in health interventions and in the avoidance of immediate environmental health disasters requires rapid assessment and response. For survivors needing care, the system of triage takes place to work out what beneficial treatment they might be offered on the basis of who, within a mixed group of people, should be prioritised.

Well-honed emergency health care can adapt rapidly to varied disaster scenarios. In the development context the health situation may already be chronic, but this fast moves to an acute state with a surge in infectious disease incidence or multiple injuries. Treating regular cases of illness in an endemic situation becomes dealing with an epidemic. One issue, however, is the extent to which model practices in health care can remain intact in emergencies. Should emergency health situations be regarded as an accentuated version of the more routine health service, or do they need to be based on a regime that demands a fundamentally different culture? There are management implications that come with emergency health, such that the situation is likely to become distinctly less inclusive and participatory. There is a shift from hard-established bottom-up principles based on primary health care to more top-down approaches, and 'experts' may be prone to taking over.

As witnessed at Goma refugee camp for Rwandans in 1994 during a cholera and dysentery outbreak, fundamental mistakes are easily made. The wrong medicines were initially prescribed by external agencies, whilst it was later found that local health posts had medical knowledge concerning what should have been used. They were, however, sidelined by the mass international relief presence at that time (Goma Epidemiology Group 1995). Selective primary health care is more justified in the emergency situation, and participation gives way to command and control. A further difference is the increase in funding that generally arrives. One of the ironies can be that during an official emergency there is less of a real health crisis in comparison with periods of 'non-crisis'. A good example would be during the floods of Mozambique in 2000, when, despite no abnormal surges in cases of diarrhoeal disease for that time of year, a substantial health intervention was realised (Collins and Lucas 2002). Oddly it was in 1998 that the real health disaster occurred, with cases of cholera at highest level ever recorded in the country and the highest that year in the world. However, in economic terms this was one of the best years for growth in the country. Many thousands of people were affected by the 1998 epidemic, but unlike in the case of floods in 2000, it was barely reported.

During conflict health services may be severely damaged. Macrae (1995) provides a detailed analysis of its impact, separating out impacts on management, organisation, economic support, its human, technical and physical resources, and disruption of health services. The physical infrastructure of hospitals is also of critical concern as it is the central component for dealing with casualties. Consequently, there have been international policy campaigns, in particular by the Pan-American Health Organisation (PAHO) to retrofit hospital buildings against environmental hazards.

Humanitarian health care practices are accompanied by a number of priority areas and minimum standards agreed by the main agencies. For example, MSF recommends ten top priorities for the emergency phase of a refugee crisis, as follows:

- initial assessment;
- measles immunization;
- water and sanitation;
- food and nutrition;
- shelter and site planning;
- health care in the emergency phase;
- control of communicable diseases and epidemics;
- public health surveillance;
- human resources and training;
- co-ordination.

Box 4.3

Sphere standards for health systems and infrastructure

1 Prioritising health services: all people have access to health services that are prioritised to address the main causes of excess mortality and morbidity.

2 Supporting national and local health systems: health services are designed to support existing health systems, structures and providers.

3 Coordination: people have access to health services that are coordinated across agencies and sectors to achieve maximum impact.

4 Primary health care: health services are based on relevant primary health care principles.

5 Clinical services: people have access to clinical services that are standardised and follow accepted protocols and guidelines.

6 Health information systems: the design and development of health services are guided by the ongoing, coordinated collection, analysis and utilisation of relevant public health data.

Source: Sphere Project 2004.

Box 4.3 shows the good practice standards that are provided by the Sphere guidelines.

The specialist requirements for emergency medical planning based more on economically advanced nations have been reviewed by Alexander (2002). Many of the issues that engage humanitarian assistance programmes in terms of policy and management are reflected in contrasting development settings.

Conclusion: integrated health and disaster management

The broad conceptual approach to this chapter has been based on a combination of supporting ideas and innovations. One framework that was referred to as a health ecology approach (Collins 1998, 2001, 2002, 2003) presented health influences as a function of six integrated types of risk. However many categories there are, health ecology and wellbeing at any one time and place are dependent on local and micro-ecologies, epidemiology, behaviour and current world context. They can be described as a shifting target requiring ongoing and integrated health assessment and management.

This disaster and development oriented approach to health addresses the legacy of deduction and overly deterministic medicalised approaches to 'combating' ill health. A focus on the hazard or vulnerability can be more possibilistic in terms of explaining the causes of disease and ill health. Consequently, integrated work on health and disease in day-to-day practice must accept a degree of uncertainty and unpredictability related to emergent health risks and human health behaviour. This is reflected in the science of identifying how pathogens and ill health afflict the poor, which is limited in part by unresolved issues relating to the nature of resurgent and emergent health risks. Advances in science and technology are, therefore, as crucial as ever to keeping pace with new and persistent health threats.

International and national commitments to health disaster risk reduction are compromised by the demands of intensive economic development. Basic issues of water, sanitation and hygiene persist. Meanwhile, marginalised and economically poor people are forced to 'risk take' to a greater extent, influencing, for example, the spread of HIV/AIDS. Despite the creation of the Millennium Development Goals, there is as yet only limited progress in devising risk reduction agendas for the existing certain health problems of poorer nations. The precautionary approach should be emphasised for macro-scale development issues such as the effects on health of climate change, chemical pollution, meat production, diet and pandemic flu. Increasing recognition of complexity and uncertainty in world health justifies a precautionary risk reduction approach consistent with current shifts in the wider disaster management paradigm. The consequences of non-investment are a lack of options for multiple early intervention strategies and limited health risk reduction. This has global significance when we consider the impact on economies due to disease uncertainty (as in the case of vCJD, SARS and avian influenza).

Engaging people in health risk reduction requires political institutions to address risk assessment and response proactively. In addressing more certain health burdens, it requires a combination of political will and location specific support that recognises value in health care at household and community level. International travel has become so intensive (particularly air travel) that pathogens can be transmitted anywhere at any time, theoretically placing everyone at risk. If we look at the case of cholera, many people live in zones harbouring *Vibrio cholerae* bacteria, but only in some locations are people sufficiently exposed and susceptible to becoming ill from them. It is therefore the focus of long-term risk reduction to understand how some people and places succumb to ill health, rather than to know how a pathogen got from one place to the next. Illness takes place when barriers between people and pathogen are inadequate, and when people and society are less resilient. It follows that transboundary health risk management is more about nation-states sharing integrated health knowledge, experience and collaborations. This emphasises multiple aspect interventions and building resilience, rather than blocking borders and spreading fear. However, the challenge here lies in communicating between differing cultures the priorities and acceptability of different behaviours in health disaster risk reduction.

Community based information systems for local risk reduction can quickly be applied to disaster response if they are already in place as a more routine risk assessment process. For example, the IDRM programme in Mozambique set up local risk committees to monitor diarrhoeal disease hazards and vulnerability in an urban setting. When an epidemic threat occurred, this network of people, already part of the community, could assist in facilitating incoming additional emergency services, in tracing risks and cases and in facilitating interventions. A local network of preparedness during non-disaster is an investment in disaster response already purchased. We can consider it an investment in certainty, rather than a reaction to uncertainty. Positively, health organisations are agreeing to principles of disaster risk reduction, as they have already been visited conceptually in the past as part of PHC. There is real opportunity that by mainstreaming health in disaster risk reduction more widely a more joined up health risk reduction programme might be seen in the coming years. This requires learning by doing Schön (1983), and the risk management process as addressed in Chapter 5. It requires that applied microbiological and epidemiological work is combined with social, economic and psychological assessments of local-level health risks. An often untapped source of information is the knowledge, perception and judgement of survivors of poverty, ill health and disasters. These contribute to integrated risk assessment as a cross-disciplinary knowledge for *early warning* and *multiple intervention* (risk management). Common principles for cross-locational risks are evident for multiple health problems, although prioritised intervention strategies may be highly variable from one place to the next. Underlying this are debates concerning priorities of investment for integrative risk reduction research and implementation, and of ethics and rights during times of uncertainty.

Discussion questions

1. Describe and contrast the broad overall changes in the distribution of ill health associated with the crises of underdevelopment with that of post-developmental societies.
2. Consider different types of infectious disease in terms of their ecological, social and economic origins. Consider the disease you identify in terms of changing pathogens, people, places, political economies and perceptions.
3. Discuss the extent to which HIV/AIDS is a pandemic disaster caused by social changes, or an opportunistic pathogen that becomes established amongst the most vulnerable.
4. Assess the extent to which famine is an outcome of food and nutrition insecurity associated with environmental change or of malnutrition caused by political economies and conflict.
5. In what way is chemical pollution the health disaster of the future?
6. Explain what may be meant by mental wellbeing in the context of development and disaster.
7. Discuss the opportunities for varying types of health intervention in terms of the responsibility of the individual and of health care institutions.
8. In what ways can the principles of primary health care be an analogy for disaster reduction more widely?
9. Consider what you think is meant by integrated health security for disaster reduction.

Further reading

Barnett, T. and Whiteside, A. (2002) *AIDS in the Twenty-First Century: disease and globalization*, Basingstoke: Palgrave.

Barrett, H. (2009) *Health and Development*, Perspectives on Development Series, London: Routledge.

Curtis, S. and Taket, A. (1996) *Health and Societies: changing perspectives*, London: Arnold.

MSF (Médicins Sans Frontières) (1997) *Refugee Health: an approach to emergency situations*, London: Macmillan.

Oxfam (1995) *The Oxfam Handbook of Development and Relief*, Oxford: Oxfam.

WHO (2007) 'A Safer Future: global public health security in the 21st century', *World Health Report*, Geneva.WHO.

Useful websites

www.doh.gov.uk/cmo/publications.htm: Department of Health (2002) Getting Ahead of the Curve (provides the basics of UK epidemic preparedness).

www.humanitarianinfo.org/iasc: Inter-Agency Standing Committee. (IASC), established in 1992 in response to General Assembly Resolution 46/182, which called for strengthened coordination of humanitarian assistance. The resolution set up the IASC as the primary mechanism for facilitating inter-agency decision making in response to complex emergencies and natural disasters. The IASC is formed by the heads of a broad range of UN and non-UN humanitarian organisations.

www.who.org: World Health Organization.

www.unicef.org: United Nations Children's Fund.

www.paho.org: Pan American Health Organization.

www.unaids.org: United Nations AIDS programme.

5 Learning and planning in disaster management

Summary

- **Planning in disaster management and development are guided by assessment, implementation, monitoring, review and evaluation.**
- **Participatory approaches ground decision making in local needs and realities, providing vulnerable people with ownership of their own solutions.**
- **Vulnerability, capacity and resilience assessment for disaster risk reduction are part of a wider set of approaches used for assessing development needs and impacts.**
- **Ongoing review, monitoring and evaluation demand that disaster reduction and sustainable development policy making are linked with practice.**
- **There is a gap between development and planning legislation and what is required for disaster risk reduction.**

Introduction

This chapter looks at the planning process in disaster management and development as a cycle of learning through doing. Probably the best understanding of disaster comes from having experienced it directly. More formalised disaster and development research involves knowledge of methodologies, methods and techniques for gathering, analysing and evaluating information about people and environments. Information gathering, learning and understanding are influenced by differences in age, gender, culture, belief, race and personality. As disasters are multi-causal and varied in nature and impact upon the knowledge required for disasters, research is invariably interdisciplinary. However, improved knowledge generation without obvious means to legislate for the policies within which it can be applied means that the effectiveness of knowledge is often either reduced or absent entirely. For example, we may know the standards of building needed to withstand earthquakes, and have the capacity to monitor this, but in reality this knowledge may be less influential than the profits that can be made from cheap to build houses. Another shortcoming in the learning process can be that circumstances during and immediately after disaster events may make detailed research inappropriate during those periods. Also, people oriented research at the best of times is subject to ethical and logistical challenges.

The process of reflection, learning and review is backed up by the participation paradigm, which has been prevalent in development work over the last thirty years or so. Participation in development can also be applied to participation in disaster reduction, the latter, for example, being now reasonably well established as 'community based disaster management'. When individuals and communities are put at the centre of development or disaster reduction strategies there are substantial synergies between the two modes of thinking. Ideally, decision making in disaster management and sustainable development would be led by demand rather than just the latest idea that happens to be in vogue. The key to responding to demand is careful assessment, which is grounded in localised and global realities verified through an integrated, participatory and scientifically sound set of processes of information gathering. This chapter concludes, however, that beyond some of the more obvious constraints of a lack of political will to act, corruption or uneven development there are also differences in learning in development and learning from disasters.

Learning cycles for disaster and development

The notion of cyclical learning in education has been well established for many years and is most frequently attributed to the 'reflective practitioner' approach (Schön 1983). Reflection without a purpose is not sufficient, the terminology therefore often being clarified as 'reflection in action'. Planning for disasters draws from the learning cycle approach and may in some respects be a euphemism for accepting catastrophe as inevitable. We work towards being able to deal with a disaster *when* it happens not *if* it happens. With disaster reduction and development initiatives potentially so closely linked, we might simply conclude that, regardless of disaster probabilities, we plan for disaster reduction through development.

One of the additional benefits of linking disaster and development is that the techniques of assessment in one are on the whole transferable to another. For example, a community based approach to development, if adopting current guidelines from development funders, will typically start with a needs assessment by the intended beneficiaries to guide the selection of development inputs (Gosling and Edwards 2003) (Figure 5.1). Similarly, a community based disaster preparedness programme might employ a vulnerability and capacity assessment (IFRC 2002; Davis 2004). Each leads to planning, implementation, monitoring, review and evaluation, alongside assessment of impact, quality and participation in programmes and other activities.

In development programming, a needs assessment includes identifying a web or tree of causal factors (a basic example was provided in Box 2.2 on pp. 79–80), which in turn define the options for reducing needs. The techniques employed to gather information, whether for disaster reduction or development, ideally include various forms of participatory activity

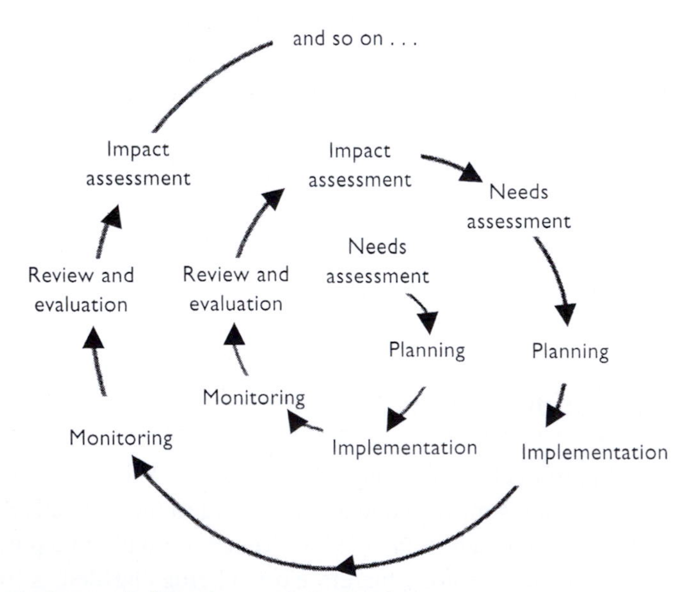

Figure 5.1 Learning cycle based on project planning

Source: Gosling and Edwards (2003: 6).

and appraisal (Chambers 1983, 1997, 2005; Pelling 2007), inductive people centred research (Robson 1993; Laws 2003) and action research (Wadsworth 1997). This can be applied to all stages of disaster planning, assessing prevention and mitigation needs both prior to and during disasters, and also response and long-term recovery post-disaster. Emergency response systems have, however, tended to rely more on an analysis of systems and procedures within the command and control structures of emergency services than on the perspectives of the wider community. One earlier version of participatory appraisal used to be termed 'rapid appraisal'. The rapid aspect in this instance also lends itself to assessment of disasters on account of the speedy information gathering that can be achieved. Rapid appraisal is what we essentially rely on for many of the situation reports (SitReps) that give information on emergency contexts around the world. However, rapid appraisal can contradict participatory appraisal if people's involvement is sacrificed for speed.

Quantitative surveys also have a role in recording a wider distribution of people and contexts of preparedness, vulnerability, resilience and so forth, such as an entire community, town, district, province or other spatial unit. Systematic surveys can subdivide cohorts of people on the basis of wealth, status, gender, age and other characteristics. There is clearly a role for both focused information gathering among small groups and through the wider survey. In practice, much of the learning required for disaster management and development can be picked up through localised knowledge, some through scientific experimentation, and all instances through ongoing monitoring and evaluation. The process of information gathering is cyclical in that once disasters have happened a 'lessons learnt' period should ensue.

Those who are caught up in the disaster undoubtedly reflect on what happened. This reflection can be formalised into a meeting of disaster response stakeholders, but ideally includes a wider group of people from the disaster and development process, including survivors. The purpose of reflection in this way may vary. In the context of emergency response and relief programmes it is to analyse what could have been avoided or executed better. Survivors' post-disaster reflection can also assist in their coming to terms with loss and mental disturbance from the event. Evaluation of how institutions dealt with the disaster typically involves the various response institutions exchanging their experiences of both what went wrong and what went right in terms of preparedness, mitigation and response. This in turn feeds into the future planning and preparedness process. Active emergency planning groups sometimes use this information for simulation exercises to improve future preparedness, including through live enactment of a response, and table-top exercises to simulate the decision making process at control centres. Live simulations require using staff and volunteers to act out events in real world surroundings, much in the way that fire personnel train to put out fires and rescue people using incidents constructed in their training centres. For the table-top exercise the simulation is based on a set of decision making processes represented by the group, for which there are known likely outcomes. As participants interact over their varying acted roles and responsibilities, the consequences of different lines of intervention can be analysed.

As suggested in earlier chapters, we are a long way off being able to precisely predict disasters and their impact. However, through the learning cycles we have mentioned, the possibility arises of reducing uncertainties and engaging in effective planning. This also contributes to early warning processes and systems, which are dealt with further in Chapter 6. Disaster knowledge, whether informal or formalised, is built up from past experiences. New information adds to the old as more disasters occur, until awareness is established and prevention and response improved. It is strengthened by local knowledge and through the documenting of disaster experience internationally. In an endeavour to capture this, and to identify good practice from various forms of disaster research, the monitoring and evaluation must engage multiple stakeholders. Whilst information is enabling, it remains largely up to the individual to administer the application of appropriate preventative action. In summary, disaster knowledge and understanding are grounded in local realities, based on personal accounts, direct involvement in disaster related activities, and accumulated assessment and review. This is what has been referred to in Disaster and Development Centre work as 'People Centred Hazard and Vulnerability Mitigation'.

Natural hazards research, as opposed to disasters research, may emphasise a people centred approach less and a technological approach more, including probabilities of physical hazard events, such as earth tremors, landslides and floods. Hazards research extends to fields such as seismology, hydrology, land form processes, geomorphology, climatology, chemistry and ecology. Though human interactions with the environment are never far away, hazards research, assessment, monitoring and evaluation in these instances are usually based on empirically grounded quantitative analyses. Hazards focused work has often been data intensive, such as with computer generated predictive models for climate change and earthquakes. However, as disasters are about combinations of people and hazardous places, the subject is ideal for those who wish to take up the challenge of integrating physical and social sciences, qualitative and quantitative studies.

Education, participation and acquisition of local knowledge

Paulo Freire's 1972 book *Pedagogy of the Oppressed* argues that education is the means to removing class differences, a liberating power, and that it must be grounded in actual experiences of students and continual shared investigation. Disaster and development education can do this by facilitating people to build awareness of contemporary threats and opportunities, but has been surprisingly absent in the curricula of most school education, save for optional tertiary level studies. Disaster education arguably empowers people to address adversity. Meanwhile, the absence of education encourages vulnerability. By not understanding the underlying origins of many types of disasters, or best practice in avoidance, mitigation and response, future generations are put at more risk. In terms of keeping up with new emergent challenges, not to be proactively keeping education forward looking and updated, would be in effect to go backwards. However, some school oriented initiatives have been gaining ground in recent years (Box 5.1).

There have been a number of studies on the nature of effective school disaster education, such as, for example, by Shiwaku *et al.* (2007) in Nepal and by Shiwaku and Shaw (2008) in Japan. Whilst the education system is an excellent way to promote disaster reduction, many people will only ever be engaged with the subject in the context of community life beyond formalised education systems (Izadkhah and Mahmood 2005). There is, therefore, also a need for a proliferation of initiatives in community based disaster reduction beyond schools.

Community based disaster reduction is driven by the increased awareness that ways forward in disaster and development work lie within the community, including through local knowledge and action facilitated by localised governance. However, many aspects of disaster management in parts of the developed and developing world remain distant from this, insisting still on more centralised systems of disaster response. Learning through action in communities is not only part of the basic preparedness strategy, but can be an empowering activity for those otherwise marginalised from access to new knowledge. Participatory approaches represent the essential procedures and also information sharing techniques that facilitate this. The aspect that Freire was emphasising, liberation through education, has subsequently been given the label 'empowerment'.

Box 5.1

Disaster reduction and schools

Education on disaster reduction may be considered relatively limited in that in some countries, such as, for example, the UK, aspects of it may only be found in parts of a single subject curriculum, such as geography, if at all. Geography itself has at times been in danger of not being considered a core topic for some secondary school pupils beyond the initial first few years. In some parts of the world, such as Japan, more attention is given to basic disaster education in schools, but largely with the motive of preparedness for individual events, such as earthquakes. The varied origins of diverse types of disaster scenarios around the world, risk reduction and conflict resolution are rarely a sufficient focus of school curricula. Meanwhile, in much of the developing world an insufficient percentage of children gain access to adequate school education in general. In response, UNICEF, UNISDR, UNDP, the Swedish International Development Agency (SIDA), Japan International Cooperation Agency (JICA) and several NGOs are supporting disaster reduction campaigns in the context of schools (Briceño 2008). Information on this is readily available on the UNICEF (www.unicef.org) and UNISDR websites (www.unisdr.org). These institutions also provide some basic learning materials for children, including a board game called 'Riskland'.
A number of school education initiatives from South and Central American countries have informed some of the materials available through this route. One example of NGO activity in this field is that of Action Aid, which has managed a 'Disaster Risk Reduction through Schools' (DRRS) project in Nepal (www.actionaid.org).

This includes not just the end result of being empowered, but all of the processes of being represented in decision making, having voices heard, and consequently a tendency for accurate and locally sensitive information. The participation of communities in risk assessment allows for the communication of information together with varying perceptions that would otherwise remain submerged. By communicating information, the risks of disasters are reduced.

Participatory approaches are therefore a route to local knowledge, empowering people and encouraging the analysis of individual's circumstances by themselves, and consequently the identification of routes to solutions. The manner in which local knowledge is transmitted to the realm of proactive decision making for disaster risk reduction and sustainable development depends on systems of governance operational at any one place and time. However, no account of the participation emphasis in development and disaster reduction would be complete without recognising some of its limitations in practice. In development planning notions of representation and inclusion may not in reality be participatory. There are varying interpretations of access and empowerment, such that attendance by people at meetings or having been consulted about a disaster or development plan might not actually be participation. People are not necessarily influencing decisions. This is a well-worked topic for scholars of participation and examples of this limitation have been documented as far back as 1969, with work originally published by Arnstein (1969) on the 'Ladder of Citizen Participation'. It indicates how at worst the participation agenda can be a manipulation, and at the other end of the ladder it can equate to citizen control.

Many of the tools for participatory approaches can be found under the broad heading of participatory learning and action (PLA) (Neefjes 2000). Where participation is true, it is about trying to see the world through the eyes of the people who are affected by whatever change is proposed (Mohan 2002). This approach is key to development planning and also to local knowledge in disaster reduction. Along the lines of Chambers' (1992: 6) version, it helps us find out about:

- complexity, diversity and proneness to risks of different communities living in varied physical environments and socio-political contexts;
- knowledge and rationality of different sectors of society, including the poorest and least powerful;
- perspectives and behaviour of different elements of a group;
- ability of local people to conduct their own analyses and solutions.

Some of the techniques used to facilitate the generation of information in participatory appraisals typically include the following:

- participatory mapping exercises;
- transect walks;
- seasonal calendars;
- activity profiles and daily routines;
- time lines;
- local oral histories;
- Venn diagrams;
- wealth ranking;
- matrices;
- inventories;
- profiling;
- folklore, songs and poetry;
- attitude surveys;

- survey of practice and belief;
- key informant, focus group and community interviews.

Additionally, information gathering may be classified as a *direct observation* exercise in which there is no specific need to use any of the above, but rather to listen, watch or engage in what is going on in day-to-day life. Traditionally, in research terms this is classified in either the 'non-participant' observation or the 'participant' observation category. However, determining the point at which time spent assessing for vulnerability, coping and resilience in the context of disasters becomes participant is very subjective. Work of this type is rare in disasters research and assessment, as people with time, resources and permission to invest in it generally end up involved in the post-disaster period, in effect observing disaster at a distance. Nonetheless, more anthropologically based development research sometimes does make a more reasonable connection with the subject matter. Some would call this approach 'ethnography'. Details on it and many other approaches relevant to data collection are available from Ellen (1984), Robson (1993) and Scheyvens

Plate 5.1 A participatory appraisal in East Timor

Source: Author.

and Storey (2003). Observation and participation in a more quantitative mode may translate to interviews, either structured, semi-structured or open-ended. These may in turn be guided by questionnaires and used in a randomised sample, or with a purposely selected group. Some people within this process can then be identified as key informants and used for extended information gathering.

Needs, vulnerability, capacity and resilience assessments

Needs analysis is a means of deciphering the array of potential deficits in people's lives that might be improved through an intervention, such as support for education, identifying a market, food security, livelihoods, water supply, shelter, health, rights and so forth. Similarly, vulnerability analysis is the identification of multiple weaknesses in human security. Each is based on a deficit approach whereby it is assumed that people are in need of some aspect of development or at risk of being part of a disaster. The assessment of capacity reveals what people have been proven to be able to achieve in relation to addressing needs and reducing vulnerability. Though aspects of the Vulnerability and Capacity Assessment (VCA) have been prevalent since the first explorations of the meaning of poverty, the tool was more formally identified in the 1990s by the International Federation of the Red Cross. VCA is a participative means of obtaining an overview of the risks and capacities existing in a country, community, household, individual or other social unit. It is similarly referred to as Participatory Vulnerability Assessment (PVA), though the inclusion of the word 'capacity' is a crucial ingredient. Capacities are the strengths people already have, which can be appreciated by looking carefully at what they are already doing. Anderson and Woodrow (1999) divide the analysis of vulnerability and capacity into the physical material (the productive resources and skills), social organisational (relations and organisations among people) and the motivational attitudinal (community view of creating change). Key principles for a successful VCA according to IFRC (2002) are as follows:

- driven by those at risk;
- full commitment;
- access to available resources;
- good training and preparation;
- participation of interested partners;
- participation of communities;

- good communication with all involved;
- a way of working.

(IFRC 2002: 143)

Three overriding principles of VCA also promoted by the IFRC are that VCA (1) puts people first, (2) is a process, not a product and (3) involves all players from the outset. (2002: 145)

A further aspect of capacity in the disaster and development context is coping. Individuals and different communities cope differently and the tendency for humans to cope is often higher than would be expected during normal times. One problem with the concept of coping in a development context, however, is that it implies that the goal is to achieve this alone. Ideally, people's expectations are to arrive at more than just survival. People have aspirations to go beyond coping, which is at a minimum threshold of wellbeing. Thus, a conversation with many of the world's poorest communities will expose children living in absolute poverty, but

Plate 5.2 A child takes part in an activity for orphans and vulnerable children (OVC) that allows expression of likes, hopes and aspirations

Source: Author.

whose aspirations are to become doctors, teachers and airline pilots, and not, for example, to merely arrive at a point of basic nutrition security.

The alternative to focusing on deficits for interventions is therefore to shift the emphasis to the assessment of existing resilience and the capacity to become more resilient. This has become a more frequently addressed theme among the NGO and academic communities since the Hyogo Agreement of 2005. This implies the need for resilience capacity assessment (RCA) rather than VCA. The question being asked in this instance is a more proactive one, as follows:

> What are the optimal theoretical, methodological, policy and practice investments that enable disaster risk reduction and development strategies to achieve sustainable human security and resilience?
>
> (DDC 2008)

This involves knowing the circumstances within which resilience is enabled and the circumstances within which it is constrained. One strategy for encouraging resilience building at the community level is through the establishment of local committees to monitor it (Box 5.2).

Key roles of the committees are:

- identification of risks;
- communication of information within community and to support groups;
- encouraging motivation of the community to reduce risks and build resilience;
- monitoring of identified risks and resilience characteristics.

The process of communication within the community about risk and resilience is key to its success. Figure 5.2 shows the idealised process whereby communication is between households, the committee and a wider support group. Established knowledge about the risks and comparison with other areas are often needed in the verification of the nature of risks, whether real or perceived ones. A stylised representation of how community based risk and resilience at the community level might sit within other governance structures is included in Figure 5.3. Whilst communities determine their own best pathways to resilience, governments have a responsibility to create a context within which these aspirations can occur. Figure 5.3 suggests that the private and NGO sectors also have roles of interacting with the community that can be appropriate to addressing risk and resilience.

Box 5.2

Risk and Resilience Committees (RRC)

The Disaster and Development Centre at Northumbria University, UK, supports the application of Risk and Resilience Committees (RRC). These have developed in the context of infectious disease risk reduction in Mozambique and in the context of multiple disaster threats in Nepal, with examination of the same in Bangladesh. They are seen as one of the natural outcomes of responding to demand for a people centred approach to disaster risk reduction. One of the key findings from this work has been that the success of the approach for any particular community group largely comes down to the motivation of key individuals in the community who wish to make it successful. Absence of this catalyst tends to mean a less successful committee. One successful committee in Beira, Mozambique, succeeded in organising the cleaning of its area without offering any payments, but leading to the reduction of diarrhoeal disease in the vicinity the following season. In Nepal, Dhankutha Municipality Risk and Resilience Committee has organised safety campaigns to reduce car accidents and a local risk register to assist the community's own monitoring of hazards and vulnerabilities. External support is of only limited value once initial training has been made accessible, perhaps most usefully in the initial capacity building for establishing these groups. Financial inputs risk detracting some groups from their purpose. Another finding has been that if linked to official structures in any way these committees need to be operational in a context of morally supportive political governance. The main functions of the committees at Beira in central Mozambique were to improve community based risk management for specific high risk locations and improve communication channels between the community and multiple stakeholders with influence in the area. A different version of these for the case of Nepal were given the role of protecting communities from vulnerability and exposure to wider hazards, informing the teaching and learning base with local knowledge, and providing a source of relevant information for circulation from community to community and to the wider governance context of the country.

Planning interventions

Assisted interventions that require more than the private actions of individuals or small groups in isolation are frequently planned within a hierarchy of policy, programming and projects. The upper level is the policy domain, reflecting the overall understanding of disaster and development, including that adopted by international organisations such as the UN, governments and NGOs. Broad scale examples are the Millennium Development Goals and Hyogo Accord, which can be broken down into smaller policy segments. Implementation of policy is then agreed through programme level strategies, within which typically the

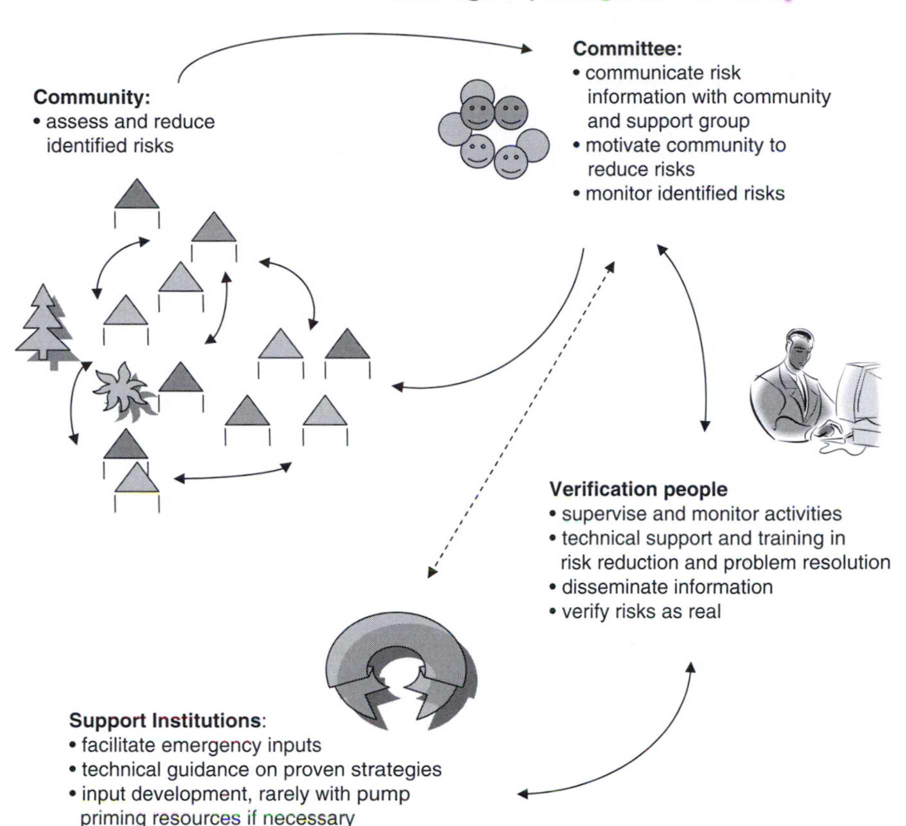

Figure 5.2 Communications within a community-based risk and resilience programme

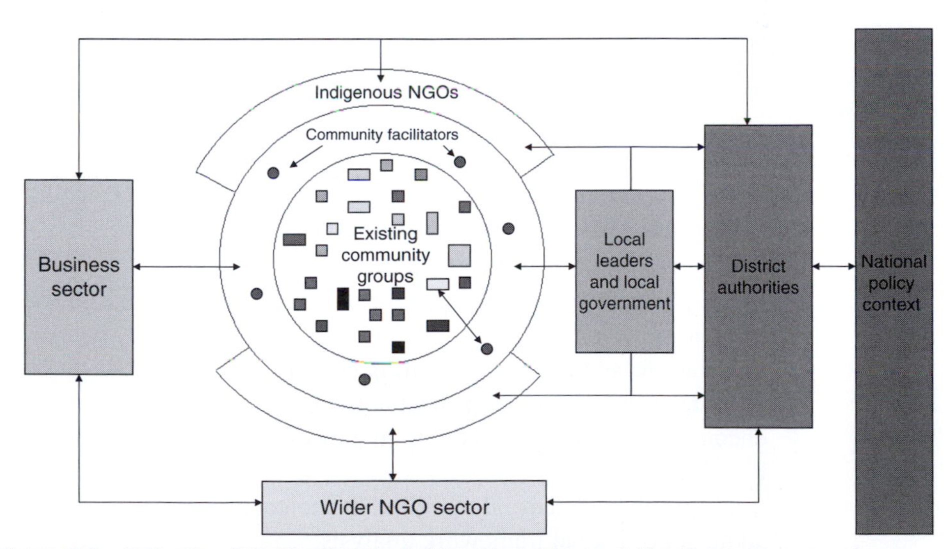

Figure 5.3 Indicative (idealised) stakeholder interactions relative to an empowered community-based organisation

main actionable policy themes are identified and budgets allocated. Programmes by international organisations to reduce HIV, or of governments to strengthen social security, or of NGOs to reduce poverty or improve rights comprise a few examples. Typically, a programme is made up of multiple projects, each of which contributes to its overall success and in so doing to wider policy development.

In order to gain funding support for a project from the public sector and many parts of the private sector, it is usually necessary to demonstrate a contribution to a programme and policy. The further requirements are that it is likely to be feasible and to make an impact. A simple tool referred to as the 'logical framework' (or logframe) has been devised to assist in structuring the planning and proposal writing process. Whilst this is by no means the only way of making a project plan, it is useful in that it relates the goal of the project to its purpose, outcomes and proposed activities. By way of example, a very summary logframe for a proposed risk and resilience committee project, similar to those mentioned above, is included in Figure 5.4. A second column includes indicators that can be monitored to demonstrate progress over the life of the project, and a third the means of verifying these indicators. A fourth column addresses the likelihood of success by identifying the risks and assumptions involved in the project. These logframes have become widely used in development and disaster reduction work at the project proposal stage. They are also referred to as a guide to project implementation and often updated during the life of the project as circumstances change. Their key function, as implied by their name, is that project activities remain linked to the project goal in a clear and logical manner. Beyond the summary table of key project parameters then represent a way of thinking in implementing development and disaster reduction initiatives. However, over the years of their application a number of strengths and weaknesses of the approach have been identified, as follows:

- Strengths of logical framework analysis:
  - links strategies, objectives and aims;
  - forces planners to consider monitoring and evaluation through use of indicators;
  - examines feasibility through stating assumptions;
  - encourages integration of project activities;
  - encourages people to consider their expectations and how these can be achieved;
  - can be easily updated and modified as project progresses.
- Weaknesses of logical framework analysis:
  - its use can become complicated;

- tries to oversimplify complexity;
- based on 'cause and effect' ideas that may be alien to some cultures and project styles;
- if based on problem analysis can be overly negative;
- can stifle creativity in project planning;
- may get too quantitative;
- prone to setting unrealistic targets;
- lends itself more to large projects.

Project Summary	Verifiable indicators	Means of verification	Important assumption
GOAL: Strengthened Disaster Risk Reduction (DRR) Governance	300,000 people contribute to Risk and Resilience Strategy by end Project Year 2	Overall evaluations of project	Government policy remains supportive of community DRR
PURPOSE: Enhanced community-based resilience and wellbeing	At least 500,000 people experience reduced risk by end Project Year 2	Baseline monitoring of community coping	Focal locations do not experience major disasters during set up
OUTPUTS: 1 Risk and Resilience Committees (RRC) established and local level risk identification and prioritisation undertaken 2 Local risk governance strengthened 3 Communities designing strategies to reduce risks 4 Knowledge of DRR enhanced amongst local development planning stakeholders 5 Creative/innovative DRR approaches shared between Southern Africa and South Asia	Minimum five RRC established in each of three sub-regions by end Project Year 2 Local government departments engaged in interests of RRC by end Project Year 2 At least three strategies identified and trialled by each RRC by end Project Year 2 Two workshops per sub-region by end Project Year 2 Fact finding visits of 10 RRC reps to other sub-regions by end Project Year 2	Audit of RRC composition and activity Participatory M+E Evaluation of governance Assessment of DRR activities Feedback monitoring of exchanges	Local compliance with the concept remains positive Political differences do not prevent risk governance Willingness of communities to implement on a voluntary basis Willingness to travel
ACTIVITIES: 1 Facilitate Community Risk and Resilience groups 2 Engage local risk and resilience governance agenda with community 3 Assist local governance structures build DRR strategies with communities 4 Train/facilitate local governance stakeholders 5 Exchange of community risk and resilience representatives and facilitating partners within South Asia and with Mozambique–includes web site, multiple guidance literature, seminars, meetings and live broadcasts	Staff time, travel, subsistence and some hardware items Mozambique £175,000 (larger country and expensive travel) Nepal £125,000 Bangladesh £130,000 Other £169,000 Total £689,000*	Audit of expenditure to activity Project monitoring and evaluation, including updating of baseline data Independent evaluation	Funders support becomes guaranteed throughout the whole life of the project

*Figures are arbitrary for the case of this example

Figure 5.4 Summary logical framework analysis for a project on risk and resilience building

Source: Disaster and Development Centre project proposal to DFID Conflict and Humanitarian Fund.

Impact assessments

Impact assessments are a range of activities that variously determine negative, positive or benign effects of changes in environment, society or the economy. There are a host of reasons why impact assessment of development is needed. Some of these are indicated in the following generalised definitions:

> A procedure for assessing the environmental implications of a decision to enact legislation, to implement policies and plans, or to initiate development projects . . . a tool in environmental management.
>
> (Wathern 1988: 3)

> Accurate, critical and, it is to be hoped, objective assessment of the likely effects of a development project, programme, policy, plan, social or economic change, environmental change etc. It should, in addition to identifying impacts, clarify what the situation would be if no development or change were to occur and what the impacts are for various possible development options. Impacts that are irreversible or that threaten organisms, environmental quality and sustainable development should be highlighted.
>
> (Barrow 1997: 299)

> The process of assessing or estimating in advance the social consequences that are likely to follow from specific policy actions or project development.
>
> (Burdge and Vanclay 1996: 59)

Environmental Impact Assessment (EIA) has been the more legally related form of impact assessment and is a well-established process in some countries. The case of the UK is illustrated in Box 5.3.

Clearly the lists in Box 5.3 only cover a narrow range of impact sources that development might bring. Some types of more subtle impacts, such as the effects on climate change of increases in traffic or household pollutants or the effects on conflict potential of economic and cultural shifts, have sources not addressed here. Within the framework of disaster risk reduction they are, however, crucial additional items of impact assessment. Social impact assessments are reasonably well established and can be differentiated from EIA in terms of the variables included. Their addition arguably brings the exercise of impact assessment closer to the context of disaster and development and impact

Box 5.3

Institutional guidance on EIA in the UK

The Town and Country Planning (Environmental Impact Assessment), England and Wales, Regulations 1999, as amended, divide the type of projects affected by Directive 85/337/EEC and Directive 97/11/EC into two categories:

- Schedule 1 lists those projects where an EIA is always considered mandatory;
- Schedule 2 lists those that may require an EIA, if they exceed certain thresholds and if they could 'significantly' effect the environment.

The local planning authority (LPA) is required to give a 'screening opinion' as to whether a development requires an EIA. The Secretary of State may also be required to give a 'screening opinion' on a development. The 'screening opinion', or 'screening direction', must be made according to the criteria set out in Schedule 3 of the 1999 Regulations. Here follows a selection of projects where an EIA is considered mandatory, as detailed in Schedule 1 of the 1999 Regulations:

- nuclear power station or reactor;
- installation for permanent storage or final disposal of radioactive waste;
- thermal power station where heat output is more than 300MW;
- an integrated chemical installation;
- gasification and liquefaction of coal or bituminous shale;
- crude oil refinery;
- extraction, processing and transformation of asbestos;
- some installations for the intensive rearing of poultry or pigs;
- a trading port;
- inland waterway capable of handling vessels over 1350 tonnes;
- aerodromes (with runways longer than 2.1 km);
- special roads such as a motorway;
- long-distance railway line.

A selection of projects in Schedule 2 to the 1999 Regulations which if above certain thresholds are considered likely to have significant impacts on the environment requiring EIA include:

- agricultural, such as fish farms, pig or poultry rearing, except where mentioned in Schedule 1, including water management on farms;
- extractive, such as extraction of coal, lignite, peat, petroleum or natural gas, ores, shale and minerals;
- energy production, thermal power station, hydro-electric schemes, wind farms transmission by overhead power cables, surface storage of fuels, production of electricity, hot water and or steam;
- metal processing activities or installations;
- glass manufacturing activities or installation;
- chemical industry activities or installation;
- food industry activities or installation;
- textile, leather, paper and wood activities or installation;
- rubber industry activities or installation;
- infrastructure projects activities or installation.

Source: Croner (2002).

assessment more widely. A list of typical SIA characteristics based on Barrow (1999: 107) includes the following:

- assessment of who benefits and who suffers – locals, region, developer, urban elites, multinational company shareholders;
- assessment of the consequences of development actions on community structure, institutions, infrastructure;
- prediction of changes in behaviour towards the affected by the various groups in a society or societies;
- prediction of changes in established social control mechanisms;
- prediction of alterations in behaviour, attitude, local norms and values, equity, psychological environment, social processes, activities;
- assessment of demographic impacts;
- assessment of whether there will be reduced or enhanced employment and other opportunities;
- prediction of alterations in mutual support patterns (coping strategies);
- assessment of mental and physical health impacts;
- gender impact assessment – a process which seeks to establish what effect development will have on gender relations in society.

Though initially applied to development contexts, the above list is arguably synonymous with what is required to assess the impact of disasters and of disaster management decision making.

Ethical questions in EIA/SIA include whether or not impacts on environments and society can be successfully costed. This is particularly poignant in the developing world, given that EIA/SIA methodologies were generally invented in the North. Criticisms of EIA/SIA highlight problems of subjectivity, devaluation through commodification and the problems of the approach being manipulated by those with interests at the hub of the development machine. However, proponents of the merits of costing the environment believe that giving it a value stops it being taken for granted, and can serve to protect it from opportunist developers. Putting a price on the environment could prevent short-termism and the hazards of plundering by those who would otherwise employ less appropriate forms of development project so as to reduce standards and therefore the costs of implementation. On the social or cultural side of the subject, it is arguable that although impacts are interpreted differently from one context to the next, there are minimum standards of wellbeing common to all humanity. The topic of standards and thresholds in development and disaster is returned to in Chapter 7.

One of the major limitations in the impact assessment approach is that although pre-project assessments are increasingly commonplace, there is often an absence of proper monitoring and evaluation of the actual impacts of projects. Compliance measures are not necessarily properly carried out within the required time-frame, absence of accurate and in-depth information in the pre-implementation and implementation phases making it easy for developers to bypass potential legislative obstacles. A further contention is whether evaluation of environmental and social impact on the basis of measurable and cost-effective benefits can ever be of universal application. There are always likely to be winners and losers in schemes which cut across wide ranging interest groups. For example, a dam project in the Amazon may be cost effective in providing an urban area with part of its electricity needs, but in the process can degrade the natural resources on which others depend. In reality, there are examples of such projects which not only have neglected the wellbeing of some groups, but have also failed to achieve their specific intended goal. Most of the cases of development induced disasters mentioned in the opening chapters of this book exemplify this. Dam projects, such as those on the Mekong, failed to make adequate EIA/SIAs as no account was taken of the cost of decommissioning, biodiversity losses and social disruption. Further to this, there was no assessment of geomorphic impacts downstream, unrealistic assessment of sedimentation and no assessment of impacts on water quality.

The process of impact assessment includes being able to identify, predict and evaluate. Stages of the processes of EIA, many of which also apply to SIA, are presented below (after Lee and George 2000):

1. Screen the important things where there is likely to be an impact.
2. Consider alternatives.
3. Describe areas of possible impact and alternatives.
4. Baseline: take measurements of items worth including in assessment.
5. Predict the impact magnitude and its significance.
6. Identify mitigation measures.
7. Document findings.
8. Review findings and establish accessible and comprehensive information for various stakeholders.
9. Consult with public and other stakeholders.
10. Decision.
11. Monitor the outcome of the decision.

12 Evaluate the outcome of the decision.
13 Review as and when circumstances change.

Impact assessment methodology can include measuring environmental and social characteristics that are quantitatively or qualitatively attributed values. Some detail on the methods that get used is provided below.

Measurement of environmental or social variables as part of scoping is carried out using a wide range of techniques, such as water and soil analysis, vegetation and demographic surveys. Some examples of methods for assessing impacts on the environment or society include the use of (1) checklists of potential impacts of refugees concentrations (Black 1998), (2) grid tests or matrices of impacts of development projects through stages of a projects development, (3) overlay analysis for spatial assessments (Wathern 1988) and (4) network analyses (or systems analysis), to assess process/feedback models:

1 Checklists are often used as a form of rapid appraisal and can be simplified to record a series of judgements as to whether a project or other impact has a positive, negative or benign influence on listed variables. It can be a simple 'tick box' approach.
2 Grid tests or matrices attempt to ascribe values to environmental impacts, so that comparison can be made between different development schemes or differences between different periods of the development. Quantification can be applied for the different stages of a project to gain an indication of its longer-term status. As such, high short-term adverse impacts may be traded off against longer-term benefits. An early, and comprehensive (in terms of its breadth) version was the Leopold matrix of 1971, in which 100 actions are set against 88 components.
3 Overlay analysis and related approaches have become technically sophisticated with the development of Geographical Information Systems. Many *coverages* (layers) of spatial information can be combined to isolate areas where impacts are maximised and minimised. This includes the use of remotely sensed data from satellites and aircraft. GISs can store, update and statistically manipulate spatially referenced information and output the results in a variety of formats. For example, digital elevation models (DEMs) can represent the product of an environmental analysis in three dimensions. The challenge for some researchers in recent years has been to try to integrate more social and economic aspects of the environment with these approaches. There are some obvious

limitations to how much can be achieved in integrating behavioural phenomena, which are often neither spatial nor easy to discern. However, there are some useful developments towards extending the GIS approach beyond the analysis of *pattern* to a more *process* oriented analysis through systems analysis.

4 Network analyses are in effect spider diagrams or systems analyses that aim to show interlinkages between cause and effect. Attempts can be made to quantify the different parts of the network.

Scenario analyses is where all of the above are carried out in a way to support answering the question 'What would happen if there is a change in influence of one or more variables?' This may be quantified in computer generated models, but in emergency preparedness this is also the field of the desktop exercise in which 'actors' work through the various possible actions for mitigating and responding to disaster impacts.

Comparison is made with other similar developments and disaster circumstances carried out elsewhere. Much of the impact assessment can be done by consultation with people who have already experienced previous development and disaster impacts. This may include specialist judgements on the basis of both formally acquired educational knowledge and accumulated or acute experience.

Attributing values to environmental information is the more controversial aspect of EIA. Methods for achieving this include contingent evaluations and environmental accounting approaches, which are often more broadly referred to as cost–benefit analysis (CBA). Typically, financial calculations determine the yearly generation of benefits and costs by either the net present value (NPV), internal rate of return (IRR), or a benefit-cost ratio (BCR) (see Dixon *et al.* 1994: 30–31 for precise definitions and example formulae). Most discussions of the problems of valuation focus on the difficulty of measuring people's willingness to pay for the *benefit* of an improved or preserved environment. However, it can be argued that the problems associated with valuing the *costs* of environmental loss and destruction are much more common and intractable. Critics of cost-benefit analysis have often drawn attention to the difficulties of quantifying environmental 'goods' and 'bads' consistently. As such, they would claim that calculation is subordinate to judgement. Some of the further challenges are as follows:

- There are spatial (i.e. downstream and 'end of pipe' environmental costs) and temporal dimensions to CBA. The transboundary impacts of development or disaster (i.e. pollution, water availability) are likely to be complex.

- There are difficulties in the transferability of CBA. Costs and benefits are relative to different locales. It is not necessarily possible to apply what has been learnt at one site to other sites.
- CBA typically uses monetary values when values can't really be costed.
- Costs and benefits are changeable concepts. Prediction of future costs is therefore not possible.
- Environmental systems on which costs and benefit calculations are based are often chaotically unpredictable. Social systems arguably need to be assessed relative to varying cultures.

However, although CBA is philosophically and politically problematic, what alternatives are there and what are the implications of not applying CBA? The alternative of an analysis vacuum may be very unhelpful in terms of highlighting disaster and development issues, such as, for example, the costs of climate change or delivering relief aid late. There may also be benefits to having some 'model analyses' as fixed reference points against which the real world can then be assessed and monitored, such that a sense of proportion and standards are maintained.

Baseline monitoring

This is used to assess the change in a project or other development activity between its start and end point and ideally also to give updates on progress towards targets at various intervals. Baseline surveys are usually based on indicators relative to different locations, components of society and different wealth brackets, the more measurable components of which are broadly represented in Figure 5.5.

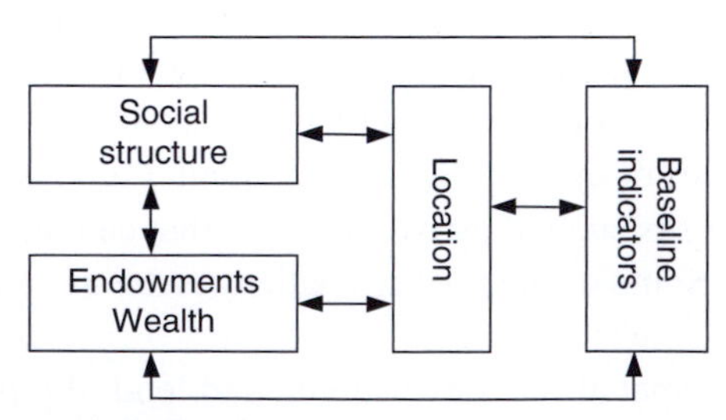

Figure 5.5 Systems approach to baseline indicators

Baseline surveys and updates have become a contractual requirement of many development interventions as a check on what changes a project may be able to identify that it has been responsible for bringing about. The methods used can include any of the quantitative or qualitative techniques already mentioned in this chapter, and will invariably be set up with costs of implementation in mind. They are typically the responsibility of the monitoring and evaluation component of projects. However, though useful for monitoring change, one of their main limitations is separating out what change has resulted from a development or disaster reduction activity from what may have been happening in any case, the so-called 'background noise'.

Risk assessment

Risk assessment is linked to impact assessment and the learning and planning cycle components mentioned in this chapter. A full account requires an examination of what we mean by 'risk'. Furthermore, risk assessment is a fundamental part of applying risk reduction and, as such, in the context of this book is also considered a part of preventative action and early warning. Beyond this, risk management is also an issue of governance such that the ownership of risk and the risk reduction processes is in itself a further vast topic that is key to disaster and development. Whilst a full section on risk assessment would also fit here as part of learning and planning, for convenience the topic is addressed as a central part of Chapter 6, which follows.

Review, monitoring and evaluation

The success of learning and planning in disaster and development is dependent on responsive review and an in-depth and inclusive monitoring and evaluation process. There are various types of evaluation that are involved in disaster and development related work. These include:

- monitoring of project impact on its targets;
- monitoring project processes;
- participatory evaluation;
- independent evaluation.

There are a number of good texts on evaluation. A simplified overview of essential points is provided by Rubin as a basic guide to evaluation for development workers, who presents reasons for doing it as follows:

> Some good reasons are to measure progress and effectiveness; to look at costs and efficient use of resources; to find out if it is necessary to change the way things are being done; and to learn from what has happened in order to make plans for the future.
>
> (Rubin 1995: 25)

A useful chapter on evaluation that can be applied to disaster and development work is provided by Gosling and Edwards (2003: Ch. 8), and a review of using evaluation to improve learning was carried out by ALNAP (2002). Cracknell (2000) takes a more institutionalised view of evaluation for the purposes of evaluating development aid. A review of different experiences of evaluation in action in the case of humanitarian assistance is provided by Wood *et al.* (2001), which is also a good commentary on how such evaluations might be improved.

Evaluation for long-term development actions has bred a somewhat different culture to that of the emergency relief field. In development, the participatory paradigm has led to increasingly widespread use of the concept of participatory monitoring and evaluation, including evaluating change in social development (Oakley *et al.* 1998) and the monitoring and evaluation of empowerment itself (Oakley and Clayton 2000). In emergency relief these concepts are frequently discussed, but there are arguably few examples of where it has been implemented. Given the pressure to respond during emergency, rather than to stop and take a lot of time to evaluate the situation, quick impact assessments have to provide the guidance required. However, evaluation of disaster relief over a longer time-frame once the emergency is under control can be routine and supportive of improved intervention, administered as a regular development activity. Evaluations are regularly carried out by the Disaster Emergencies Committee (DEC), which coordinates the emergency funds destined for a wide range of humanitarian agencies and NGO relief implementers. The evaluation of humanitarian aid is covered for seven countries by Middleton and O'Keefe (1998). A further evaluative critique of disaster response is found in Anderson (1999) and Wood *et al.* (2001). However, most of the evaluative materials on disaster response remain as 'grey literature' (reports) filed by humanitarian and consultancy agencies. There are several routes to these, including through direct contact with relief agencies and their websites, and through coordinating bodies such as (DEC) and the Active Learning Network for Accountability and Performance in Humanitarian Action (ALNAP).

The monitoring and evaluation process requires criteria and SMART indicators (Oakley *et al*. 1998; Gosling and Edwards 2003). SMART stands for Specific, Measurable, Achievable, Relevant, Time-bound. Widely used evaluation criteria include the following:

- efficiency;
- effectiveness;
- impact;
- relevance;
- sustainability.

However, the OECD Development Assistance Committee (DAC) evaluation criteria extend this somewhat. They indicate that 'efficiency' measures the outputs in relation to inputs, that 'cost-effectiveness' looks at alternatives to see whether more efficient processes might lead to greater impacts, and 'effectiveness' at the extent to which the activity achieves its purpose. 'Impact' looks at the wider effects of the activity, 'relevance' at whether it is in line with local needs and policy, as well as donor policy, and 'appropriateness' at the extent to which activities are sensitive to local contexts, ownership and needs. 'Sustainability' assesses the long-term viability of the activity's impact, 'connectedness' whether it takes longer-term interrelationships into account, 'coverage' whether it reaches major populations, and 'coherence' checks that there is consistency of representation and activities (ALNAP 2002).

With all of the above information processing there remain limitations. This is largely in the interpretation of information provided, such that it is important to reflect on the following questions on additional influences before any more final judgements or major decisions are made:

- Who is doing the interpretation?
- Who decides what is beneficial?
- What are the trade-offs in acting on different types of information?
- How do we identify the poorest of the poor/displaced/vulnerable? This depends on definitions of poverty/displacement/vulnerability.
- What information should be used to guide project implementation?
- Is it safe to generalise? For example, on how different types of households choose to behave.

Furthermore, good practice in disaster prevention and development work needs more than just good planning, review, monitoring and evaluation. In this era of raised awareness about the benefits of reducing the risks of disasters before they occur, there is an opportunity for greater and effective legislation to be put in place. This is to demand that potentially negative impacts on people, including unnecessary additional climate changes, are avoided.

Conclusion: linking demand led practice to policy

All aspects of disaster and development work are part of an ongoing process of learning how to address the demand for improved human wellbeing in continuously changing environmental, social and economic conditions. This requires regular assessment of needs, vulnerability, capacity, coping and resilience, to facilitate planning and implementation, which in turn is improved by monitoring, evaluation, impact assessment and review. There are multiple quantitative and qualitative information gathering and dissemination techniques available to facilitate this process, many of which are similar to those used by the wider research community. Researching disasters involves a similar range of methods that are required for researching development, either line of inquiry

Table 5.1 ***Learning in disaster, development and disaster risk reduction***

Learning through disasters:	***Learning through development:***
• Time is compressed • Years of research can be replaced by finding something out in an extreme moment • Mistakes are made – hasty and inappropriate reactions – misrepresentation of nature of cause • Learning is accelerated – through direct involvement and through post disaster reconstruction and evaluation • Actions are more reactive	• Long time frame • Multiple cross-comparisons of experiences • Mistakes are re-evaluated through different development approaches • Learning can be very slow but well informed • Actions tend to become widespread and reflective of established principles

Characteristics of people centred risk management education:

- Issue driven – demand led
- Practitioner oriented – demand interpreted
- Risk assessment – multidisciplinary and integrated
- Learning by doing – includes versions of risk management cycle
- Lessons learnt – reflection in action
- Local knowledge – inductive and grounded research
- Participation – people centred
- Proactive engagement – with hazards, vulnerability, coping, resilience
- Support is to date limited

demanding a participatory approach if the intention is better understanding of real demands and improved decision making. However, some differences between development learning and disasters learning are worth noting. Table 5.1 summarises a few of them. With learning comes the responsibility for action. The current increase in interest in disaster studies has improved much of the development learning, of which this book is a small part. However, there is still a very long way to go in managing to achieve the level of action that is required. Whilst technically, and in terms of representation, they are essential, one risk of the 'participation' and 'community based' approaches in both disaster reduction and development is that government may feel their support is not needed. However, a disengaged government is potentially a hazard to this agenda.

Discussion questions

1. How can disaster and development work be guided by the principles of reflective practice?
2. In relation to an extreme disaster or development issue, compare and contrast the likely differences in applying needs, vulnerability, capacity, coping or resilience assessments.
3. What are the benefits and constraints of detailed planning of disaster and development interventions?
4. How are impact assessments achieved, and what are their roles in implementing disaster and development projects?
5. Discuss what new evaluation criteria disaster risk reduction might need to address.
6. Is it inevitable that learning from development should be separate from learning in disaster?
7. In what way should governments remain engaged with their people concerning disaster risk reduction?

Further reading

Birkmann, J. (ed.) (2006) *Measuring Vulnerability to Natural Hazards: towards disaster resilient societies*, Tokyo: United Nations University Press.

Cracknell, B.E. (2000) *Evaluating Development Aid: issues, problems and solutions*, London: Sage.

Estrella, M. (ed.) (2000) *Learning from Change: issues and experiences in participatory monitoring and evaluation*, London: ITP.

IFRC (1996) *Reducing Risk: participatory learning activities for disaster mitigation in Southern Africa*, Geneva: IFRC.

Lewis, J. (1999) *Development in Disaster-prone Places: studies of vulnerability*, London: ITP.

Rubin, F. (1995) *A Basic Guide to Evaluation for Development*, Oxford: Oxfam.

Twigg, J. (2007) 'Characteristics of a disaster resilient community'. Version 1, for the DFID Disaster Risk Reduction Interagency Coordination Group, available at http://www.benfieldhrc.org/disaster_studies/projects/communitydrrindicators/community_drr_indicators_index.htm

Useful websites

www.iema.net: the Institute of Environmental Management and Assessment (IEMA).

www.alnap.org: Active Learning Network for Accountability and Performance in Humanitarian Action (ALNAP).

www.mande.co.uk/news/htm: a news service focusing on developments in monitoring and evaluation.

www.ids.sussex.ac.uk/ids/particip/: the Participation Group at the Institute of Development Studies, Sussex University.

www.iied.org/resource/: the Centre for Participatory Learning and Action at the International Institute for Environment and Development.

6 Disaster early warning and risk management

Summary

- **Predictive models, early warning and risk assessment variously help to identify the likelihood of disaster.**
- **A wide range of socio-economic, behavioural and physical research techniques can be employed as part of disaster early warning systems.**
- **Institutions that engage with early warning are part of elaborate information system infrastructures.**
- **Unpredictability is an inherent aspect of disaster and development requiring subjective risk assessment and decision making under conditions of uncertainty.**
- **Disaster risk reduction is a governance issue and can be managed as part of the development process with benefits for safety both now and in the future.**

Introduction

A straightforward logic understood by those who have undergone crisis is that information in advance can enhance the chances of avoiding or reducing the impact of the event. If disasters are considered unnatural, brought about by being in the wrong place at the wrong time without adequate protection, early warning is the knowledge about where not to be, when, and what one should do to get protected. This knowledge is based on personal experience, access to empirical secondary data, and awareness and understanding of risk. The communication of risk depends not only on good data, but on an adequate way of transforming data into accessible information, and on the adoption of its meaning by the recipient. Early warning is therefore an aspect of disaster management dependent on the development of human awareness and the capacity to decipher information, knowledge and adapt appropriate reactions to risks and warnings. In this sense, all people are part of an early warning system regardless of the formalised development of this field. This is important, not least because human responses to varying risk thresholds are very variable. Early warnings are merely the communication of information for which those listening must then engage individualised risk assessment and reactions. As we have seen from the earlier chapters, what we have described as disaster phenomena are invariably the outcome of multiple risk influences. Arguably, no two people interpret risks in an identical fashion. Risk assessment is a function of the perception of threats and rewards, past experience, received knowledge, economic constraints, environment, personality or genetic predisposition.

This chapter provides a basic account of the development of early warning in disaster prevention and of the application and implications of risk assessment to risk reduction. The range of approaches and methods used in this aspect of disaster and development are diverse, including economic, social and environmental approaches, use of technology or otherwise. Whilst the powers of prediction are routinely engaged by disaster risk analysts, in reality uncertainty prevails across the disaster and development nexus. However, despite much realisation of uncertain futures for humankind, significant progress is being made in the art of reducing uncertainty for improved chances of sustainable futures. The practicalities of effective early warning systems contrast somewhat from in-depth risk dialogues found in the literature, but are in practice inseparable parts of the same equation.

Early warning systems for some of the world's greatest disaster events have proven to fail when they are most needed. Early warning fails either because correct information is not sent, correct information is not understood by the recipient, or because of choices not to believe or heed the warning. Particularly in the case of slow onset disasters, such as famine in the Horn of Africa, we have witnessed how information may be prevalent but the political will for timely intervention absent. In the case of rapid onset emergencies such as the South Asian Tsunami, proliferating communication technology that development has brought to the region actually failed to deliver what was required in the hour of need.

There was evidence that where indigenous knowledge transfer was more intact protective actions were achieved, but that basic knowledge was in most locations found to be absent from the memories of the coastal populations. We can conclude here that good early warning systems are a part of essential developments in disaster reduction, but that ultimately this depends on the knowledge, capacity and motivation of recipients of information to act on it.

Interpretations of risk and risk management are also therefore about dealing with real and perceived risks, risk decision making and all those factors that intervene in this process. One of the greatest challenges facing the implementation of disaster risk reduction is how to engage analyses, actions and review during periods when no obvious disaster is upon us, as the essence of prevention and preparedness during periods of non-crisis. It is here again that we face the challenge of bringing development into disaster reduction and disaster reduction into development.

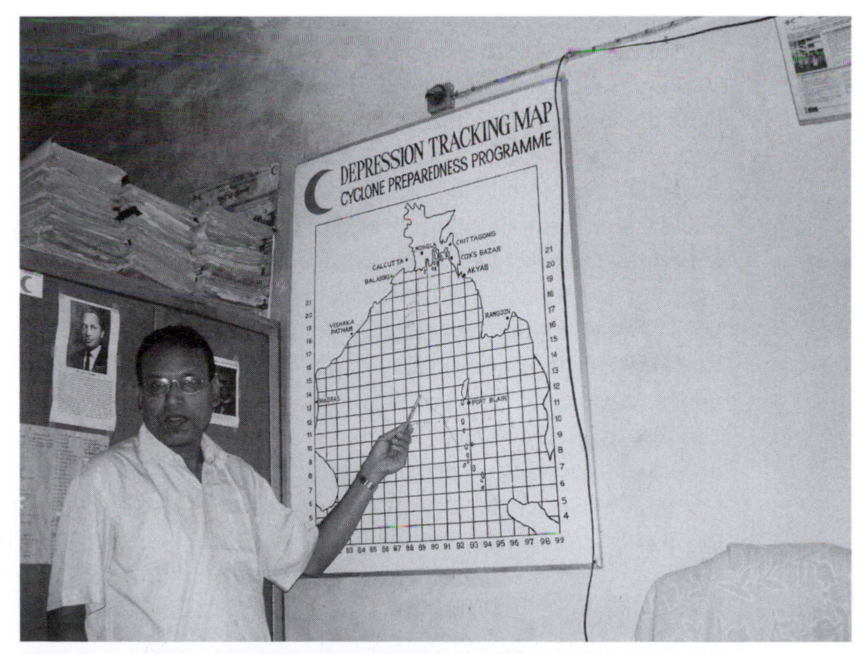

Plate 6.1 Inside one of the many communication centres of the Cyclone Preparedness Programme in Bangladesh

Source: Author.

Hazard, vulnerability and resilience indicators and thresholds as components of early warning

The essence of early warning is to monitor and communicate an analysis of indicators of change. Protection from an oncoming disaster event means adapting to the knowledge of changing indicators. This was realised long ago by those engaged in finding ways of avoiding famine. The study of early warning was developed in a more concerted way in this context. As famines can be slow or more rapid onset, they provide the ideal first case study in this instance. Devereux (1993) identified a diverse range of theories of famine, the art of effective interpretations and communication being key to adequate early warning (Buchanan-Smith and Davies 1995). An early warning system is not only dependent on the effective communication of information, but on having an adequate set of indicators for each hazard and vulnerability type.

The systems approach to famine early warning is neatly examined by Walker (1983). It is central to identifying the stakeholders and gatekeepers in the process of communication of information and associated activities. However, the political constraints to this in complex emergencies is part of what lies behind some of the monumental failures to avert disasters in recent times (De Waal 1997). Famines may be the most predictable of disaster events. Most have been foreseen early on. Varying indicators have been exposed by various disciplinary lines of thought, also requiring that famine early warning be seen as part of a wider 'food security information system' (Devereux and Maxwell 2001). Famously, Sen (1981) highlighted famine as a function of production, trade, labour and inheritance based entitlements being gradually reduced until total collapse occurs, influences that can be monitored as early indicators of the onset of food insecurity. Economic indicators, such as price changes and market failure, are developmental warnings of this. Risk of drought or destruction of food production through events such as floods, storms and pests requires further sets of indicators. They would also ideally include preparedness for emergency relief and how that can impact on food crises when local trading networks are not sufficiently robust. Slow onset indicators might be based on crop failure leading to increased food demand, declining sale of food, hoarding, a rise in food price and the onset of increased impoverishment amongst the poorest households.

Monitoring indicators can be complicated if the full set of influences on a local food security system are considered. Hubbard (1995) showed how they are a function of the food market, work and income, and micro-level household factors, but also links these to the disease environment of an area, fuel and water markets, as all of these impact on nutrition. MSF (1997) list the following indicators of food insecurity for the contexts of mass emergencies where refugee movements are likely to occur:

1 Early indicators:
   - rains late or failed;
   - crop failure;
   - deteriorating economic situation;
   - increased grain prices;
   - unseasonable disappearance of foods;
   - low levels of household stores.
2 Stress indicators:
   - increased dependence on wild foods;
   - increased dependence on food aid;
   - decreased number of meals.
3 Outcome indicators:
   - increased malnutrition;
   - increased mortality;
   - outmigration.

A wide range of indicators can make up famine and other early warning systems including indicators of social stability, prices and what can be observed regarding changes in environmental, livelihood or health security. Monitoring of environmental indicators can range from gathering remotely sensed data from the earth's surface via satellite to entitlement and behavioural mapping indices. Combinations of quantitative and qualitative information are required to address hazard, vulnerability and associated statuses using a wider set of indicators. Effective communications based on changes in multiple risk types are essentially the core ingredients of integrated early warning systems. The ultimate goal is to convert observed indicator data to information that can be synthesised into knowledge for decision making and implementation of avoidance, mitigation and response strategies. If we extend our thoughts to, for example, preparedness for a nuclear war, early warning systems have also concerned monitoring preparedness for long-term recovery after the predicted catastrophe has taken place.

Similar principles that developed around famine early warning apply to warning about other major catastrophes. For example, early warning about a disease epidemic might be based on temporal changes in the distribution of cases compared with known outcomes related to previous data. The build-up of information on seasonal cycles and disaster phenomena allows for prediction in either simple or complex models. By monitoring the changing vulnerability of susceptible communities, also in relation to environmental changes, it is possible to gain an impression of likely forthcoming epidemic threats. The health system development context within which this can be set up as a formalised process was addressed in Chapter 4. A further example of early warning in action would be the case of monitoring unrest within communities alongside the emergence of bad governance and revolt. Taken together with indicators of extreme deprivations of rights and of poverty, we may find 'a recipe for disaster'. Early warning of this type, together with sufficient political will to take action to defuse the situation, is what would be meant by conflict risk reduction.

Much of early warning is developmental over long time-frames. However, systems for alerting people to disasters are more commonly associated with rapid onset emergencies. Smoke and fire alarms go off,

Plate 6.2 Monitoring ecological early warning indicators, Bangladesh. Changes in algae in this pond may be associated with heightened likelihood of cholera
Source: Author.

Plate 6.3 Monitoring food security through a household-based interview in Mozambique

Source: Author.

loud hailers are used to call out to communities, radio broadcasts transmitted and information sent from institution to institution to alert multiple emergency response roles. The weather forecast is one of the main sources of advance information concerning impending meteorological crises. For example, throughout the Caribbean, Central America and USA, in particular, meteorological offices are on the alert for hurricanes. These are given individual identities, such as Hurricane Mitch in 1998, Michelle in 2001, Ivan in 2004, Katrina in 2006, and many others. The IFRC in 2005 identified Cuba as not only one of the most hurricane prone areas of the world, but also as one of the most prepared. It is reported that when Hurricane Charley hit Cuba on 13 August 2004 70,000 houses were damaged but only four people died (IFRC 2005).

When Hurricane Ivan got close to one end of the island later in 2004 over two million people were evacuated and nobody lost their life. To be able to execute such a plan demonstrates the need for early warning that is linked to preparedness. Meanwhile Hurricane Ivan devastated Grenada due to the government's delay in calling for preparedness measures. Hurricane Mitch in Honduras also had a contrasting outcome. Lack of any evacuation procedure by way of preparing communities, together with the physical vulnerability of settlements, led to a large loss of life. When Hurricane Katrina hit the Southern United States in August 2005 it severely impacted on New Orleans, considered one of the most vulnerable cities to flooding in North America. Much of what survived the initial hurricane damage then succumbed to extensive flooding caused by the rupturing of defensive dykes under the stress of high tides and excessive storms. Warning of an environmental hazard is rarely sufficient to predict all of the potential permutations of risks that might apply, though much of the Katrina consequences had been predicted by disaster analysts who had worked in that region.

Warnings are sometimes missed altogether, as was famously the case with the United Kingdom's greatest storm, which occurred on 16 October 1987, the worst to affect Southeast England since 1703. This was caused by an intense climatic depression (960mb) off the northwest coast of Cornwall. It moved in rapidly and weather forecasters were unable to predict its track and ferocity, though severe warnings were flashed to emergency services once it was realised what was going on. It is estimated that 15 million trees were felled as a result and there was widespread damage, though the death toll was held at 16 people. Many of the examples suggest that the early warning of a crisis, beyond monitoring hazard events, must assess the vulnerability conditions that will be confronted. The impact of an event is mitigated by preparedness, which is a function of having the right information, political will, capacity and resources amongst other things.

Given that early warning is not just about data, information and knowledge, but the capacity and political will to implement a strategy, it is important to reflect on why warnings are sometimes ignored. Some explanations have been outlined along the following lines in an IDS briefing paper (IDS 1995: 2):

- Bureaucratic procedures are not geared to the needs of recipients. Decision-making is too slow, and delays in delivery mean that assistance arrives too late.
- Decision makers only take action when there is clear quantitative evidence that a crisis is already underway – qualitative indicators are ignored.
- Early warning information is distorted to suit the needs of interest groups, who may play down or exaggerate the threat, making relief agencies sceptical of information.

Box 6.1

Cyclone Preparedness Programme (CPP)-Bangladesh

The Bangladesh Red Crescent Society is operating the CPP on behalf of the Government of Bangladesh through the Ministry of Disaster Management and Food. The idea of the CPP programme dates back to the 1950s, but was given added meaning and momentum after 12 November 1970, when a major cyclone hit the coastal belt of Bangladesh with a wind velocity of 62 metres per second. There was an accompanying storm surge of 6–9 metres in height, killing an estimated 500,000 people and making millions homeless and severely destitute. In the 1990s, Bangladesh absorbed a further five major cyclones. Out of these, most of an estimated total of 140,000 people died during the great storm of 1991, but about 350,000 people were successfully evacuated. In 1997, fewer than 200 people died, whilst about 1 million were evacuated. Overall, an estimated 2.5 million were safely evacuated from likely death in the 1990s and this was thanks to the CPP. The programme can now alert about 8 million people across the coastal belt and assist about 4 million of these to be evacuated. There are 143 radio stations that can relay messages to 33,000 village volunteers who use megaphones and hand operated sirens. The programme has managed to:

1 disseminate cyclone warning signals to local residents;
2 assist people in taking shelter;
3 rescue victims affected by a cyclone;
4 provide first aid to people injured by a cyclone.

It is estimated that the annual operating expense of the CPP in 2001 was US$460,000 (government: 56 per cent; IFRC: 44 per cent). Construction costs of a cyclone shelter were approximately US$78,000, with annual running costs at US$780. Further uses of the shelters include as schools, mosques, clinics and centres for extension work.

The achievements of the CPP, and those who have supported it, have clearly been monumental. However, as with all disaster early warning and preparedness programmes some limitations inevitably persist. One DDC project listened to how in 2007, although people received warnings about cyclones, flash floods were often missed, catching villagers off guard. A village volunteer, who acted as spokesperson, remained very poor and indicated not having batteries for his megaphone, and no protective clothing. Some community members strongly hinted that there was only limited recognition for the bravery of these key people in saving the lives of others. In some instances, the organisation of a cyclone shelter is felt to present its own problem in that the person selected to manage it may not live immediately by the shelter. There was therefore a delay in opening a key shelter up during the threat of Cyclone Sidr in 2007. Women are known to have been more vulnerable during several of these cyclone events due to a greater tendency to be in the home during the crisis, being caught up by their traditional clothing and not being able to swim, and in some instances being reluctant to move into the cyclone shelter for reasons of privacy (further analysis of the impact of disaster on different groups is made in Chapter 7). As the system is area based, those travelling long distances from one area to another for trading purposes died away from home and therefore did not appear on lists of families who would receive assistance. The issue of the increasing frequency and impact of flash floods was also referred to in the context of changing land uses along the coast. There are very obvious ecological impacts from intensive shrimp and salt farming, with wide areas bare of vegetation and exposed to the onslaught of the tides. A proximate and underlying uncertainty pervaded the Bangladesh coastal area in the light of knowledge about a likely forthcoming sea level rise and increased storms with climate change. However, the tenacity and organisation of those who live in this region have demonstrated some of the finest characteristics of resilient communities known.

Source: IFRC (2002) and author personal communication in this region.

- Warnings may be ignored altogether if political relations between the necessary collaborating government or non-governmental entities are strained.

Approaches and techniques for information gathering in early warning and risk assessment

Reference is also made to the learning techniques outlined in Chapter 5. The key to good surveillance for early warning and other aspects of disaster mitigation planning is the understanding of the nature of influences that increase crises risks. Much of the integrated approach to disaster and development is relevant here. We have learnt that knowing the status of disaster threats requires information gathering and understanding of both environmental hazards and human vulnerability at both macro-level and micro-level. This also becomes an active process of learning lessons through assessing capacity in interventions, finding out what worked and what did not last time around. VCA and resilience monitoring were referred to in Chapter 5, which emphasised how risk assessment can be embedded and owned locally so that individuals or whole communities engage in interventions.

Some of the methods of early warning information gathering techniques also include a number of specific systems and techniques for forecasting. The modelling approach is addressed by Alexander (2002) as a fundamental of emergency planning and management. Attempts are frequently made to model geophysical processes leading to earthquakes, climate change and related catastrophes. However, it is worth noting that despite decades of scientific and mathematically based modelling, earthquake science has only provided us with the ability to predict earthquakes very broadly. Geological evidence suggests about a 300-year interval between mega-quakes off the coast of Cascadia in northwest North America. Since the last one occured in the 1700s, the probability that another will occur soon is increased. It would decimate the city of Vancouver, but no one can know when it might actually occur. If predictions are only broadly correct, then the city of Istanbul is also currently in a period of higher earthquake risk.

Discussion of prediction and uncertainty continues, with a brief comment on short-term and long-term predictive models. A greater time-frame for knowledge about past events can enable better prediction. Statistical modelling can be very advanced, but with more modest statistical ability users of multiple regression can apply it so long as data are sufficiently accurate. However, the manipulation of numeric data is only one part of the modelling process, which requires input of expert knowledge if it is to be interpreted sufficiently. Most modelling of environmental phenomena requires non-linear approaches and interpretation. The success of both predictive modelling and early warning is dependent on the quality of information that can be put into the system prior to its analysis. The struggle for good sources of data is in turn based on in-depth understanding of what is relevant information. This is dependent on the techniques and skills across a range of environmental, social, economic and integrated surveillance techniques.

An example of a macro-scale physical surveillance contribution includes the United States Agency for International Development (USAID) Famine Early Warning System (FEWS), which has its origins in the 1970s working in collaboration with the FAO to predict crop yields. These sources of information are often combined with other information for situation reports for which there is open access via ReliefWeb (www2.reliefweb.int). GIS has facilitated more sophisticated spatial assessments than would have been possible in the past. In its simplest form it is a system of digital maps and database storage and retrieval. However, this tool of information technology has the capacity to also be a significant contributor to some aspect of spatial hazard and vulnerability analysis. For example, it allows multiple layers of information to be built up linked to process modelling for a distributed risk map, as schematically represented for the case of cholera risk assessment (Collins 1996) (Figure 6.1). Combinations of grid squares can be superimposed on each other to calculate spatial areas of higher or lower risk. The smaller the square, the finer the resolution will be. Environmental assessments include specific techniques such as assessing the stability of hill slopes, avalanche surveys, environmental health risk assessments, and any of the environmental management methods, tools and techniques for information gathering. A chapter dedicated to environmental monitoring techniques is provided by Barrow (2005) in *Environmental Management and Development,* included in this series.

Further quantitative techniques for gathering the information needed for early warning and risk assessment include baseline surveys that assess change from a cross-section of conditions at previous times, food balance models that monitor inputs and outputs in food supply and demand frameworks, and market behaviour models. Anthropometric surveillance, the measurement of the human body, is usually reserved for managing nutritional crisis once it has already arrived. For more qualitative

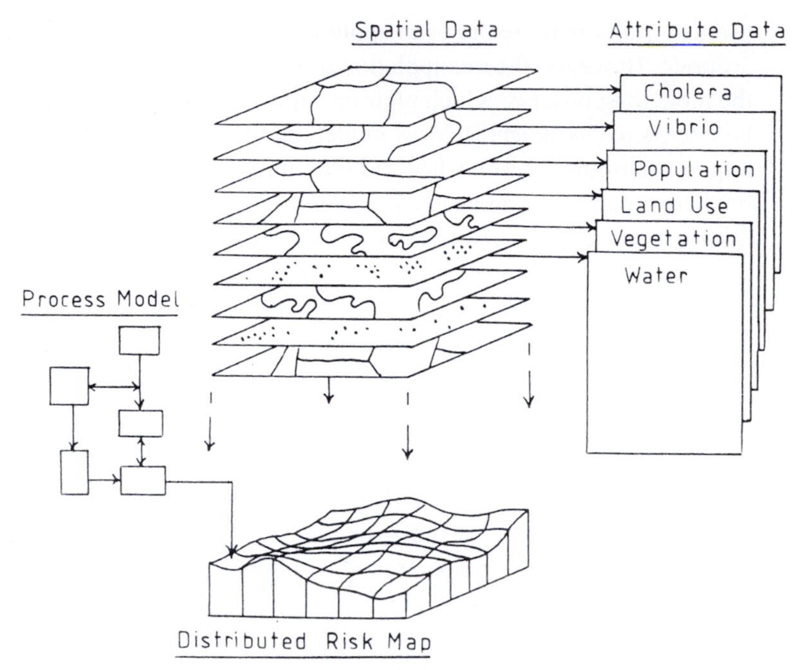

Figure 6.1 Geographic Information System (GIS) overlay analysis producing a risk map
Source: Collins (1996: 285), with kind permission of Springer Science and Business Media.

information the entire section of Chapter 5 of this book addressing participatory techniques as part of monitoring development is highly relevant here (see pp.181–188). The tools and techniques of social research and monitoring in development are the same as those required for disaster early warning and risk assessment. Entitlement mapping has been mentioned already in the context of early warning of food insecurity, but along with stakeholder and political analyses is key to the forewarning of social, economic and environmental crises. As the world has woken up to the realisation that disaster risk reduction is not just about avoiding inevitable environmental hazards, but dealing with social, economic and rights based determinants of vulnerability, such techniques are arguably the new necessity of early warning systems and risk assessment information gathering.

Risk and uncertainty

This section provides an applied account of key points about risk and uncertainty of central interest in disaster reduction and sustainable development. Whilst risk and uncertainty are fascinating topics that could be examined in mathematical and philosophical ways, their importance to this book is first and foremost in line with what Twigg (2007b) has referred to as a 'common sense approach' to disaster terminology. The issues for a disaster and development approach presented by work on risk and uncertainly are extensive, a few of which are as follows:

- the extent to which risk can be quantified;
- the implications of actual, perceived and constructed risks;
- how to live with unpredictability and uncertainty;
- the economic, social and environmental costs of uncertainty;
- risk transfer and risk governance – the rights and responsibilities issues.

By way of introduction, it is reasonable to view risk as a function of uncertainty. Although there is perhaps no universally accepted definition of risk, the normative science based sense is where odds can be calculated. The quantitative approach to risk is therefore often represented as a series of probabilities. For example, if 50 people out of 600 get TB in a population of at-risk people, the risk of TB among the 600 is 50/600, which gives a probability value of $p = 0.08$, or an 8 per cent chance. Under such conditions we can see why traditional definitions of risk have been along the lines of:

$$\text{Risk} = \text{Probability of an event} \times \text{Magnitude of the event}$$

Another version of quantitative disaster risk based on loss mitigation where probability is more or less known is as follows:

$$\text{Risk} = \frac{\text{Hazard (probability)} \times \text{Loss (expected)}}{\text{Preparedness (loss mitigation)}}$$

(K. Smith 2001)

However, uncertainty also extends to where probabilities are not known. We consequently arrive at two types of uncertainty, one which is probabilistic and theoretically measurable and another that is not, instead being governed by individual subjective judgement and structural constraints. Non-probabilistic uncertainty is where information is simply missing or is not assumed to be sufficiently accurate to calculate probabilities. Recent studies in homeland security and biological conservation have applied info-gap decision theory (Ben-Haim 2006) to investigate the issue of economic costs associated with unknown possibilities.

Uncertainty is also where complex cultural, behavioural and political-economic interpretations render risk a construction of ideas and therefore the function of a more perceptual world than an actual world. The tension between subjective and objective risk taking is one of the core issues that is examined in depth by Adams (1995). The cultural construction of risk was addressed by Douglas (1986), and the theme of how modern society has become driven by the increasing risk context within which it develops is a main theme engaged by Beck (1992). Models of macro-scale environmental-human risk interactions, together with a collection of environmentally based case studies, are provided in the edited volume by Kasperson and Kasperson (2001). It is clear within each approach that there are combinations of voluntary and involuntary risks, the latter has implications for whether people have a legal right to take risks that threaten their own wellbeing, but more importantly that of others.

For the purposes of bringing disaster and development issues together, this book has frequently addressed disaster as a function of both hazard events and, in particular, changing vulnerability and resilience. Therefore, the core idea of Risk = Hazard × Vulnerability (Blaike *et al.* 1994; Wisner *et al.* 2004) is more appropriate. This is particularly the case given that 'capacity' as mitigating disaster has also developed from this approach, such that:

$$\text{Risk} = \frac{\text{Hazard} \times \text{Vulnerability}}{\text{Capacity}}$$

The disaster and development confluence can be made to fit this last equation in a straightforward way if development is considered in its more positive light as akin to capacity. However, again in the context of this book, it would be reasonable to simplify the theoretical generalisation to:

$$\text{Disaster risk} = \frac{\text{Unsustainable development}}{\text{Sustainable development}}$$

and

$$\text{Development risk} = \frac{\text{Disaster risk}}{\text{Sustainable development}}$$

or just

$$\text{Disaster risk reduction} = \text{Sustainable development}$$

Where sustainable development includes the propensity for human and environmental systems to be not just vulnerable or more resilient, but actively engaged towards the wellbeing of present and future generations in all possible manners.

Though this can come across as grandiose and idealistic, the logic has application to practice in risk reduction work. The current resilience paradigm approach would appear to be pointing in this direction, but it would take either desperation or concerted work of decades to alter current human behaviour to take on the full meaning of risk reduction through a wellbeing approach. Decision making where there is maximum uncertainty can be guided by moral, economic or other imperatives, and world history has been decided on those grounds rather than calculation of risk. Important principles come into play, one of which, in particular, is referred to as the 'precautionary principle' in environmental management and neighbouring disciplines. As with most of the terminology, it means different things to different people. One interpretation would be that precaution under conditions of uncertainty is too simple not to 'go there' or 'do it'. Therefore, if you are unsure about the extent of further pollution of the atmosphere bringing climate catastrophe, then just stop polluting. However, the reality is quite different.

The approach advocated by the European Environment Agency (EEA) in *The Precautionary Principle in the 20th Century* (Harremoës *et al.* 2002) would appear to emphasise the precaution as the eternal quest for more information to try to reduce uncertainty. Clarification of uncertainty and precaution has been suggested by the EEA as needing to be separated out into: 'risk', where preventative actions are taken to reduce known risks, 'uncertainty', where precautionary prevention is action to reduce potential hazards; and 'ignorance', where action is taken to anticipate, identify and reduce the risk of 'surprises' (Harremoës *et al.* 2002: 217). Harremoës *et al.* conclude with a list of twelve late lessons from early warnings as presented in Box 6.2, but also a statement that action along these lines requires a profound ethical and political commitment.

Disaster risk reduction and governance

Some of the broad pathways possible in the face of an unstable disaster threat are outlined in Figure 6.2. In some idealised world, reduced uncertainty about the event might equate to reduced instability and impact of the disaster threat. What we know, we can act on to avoid a catastrophe. Risk assessment includes identification of a risk, its estimated impact, social consequences of derived risk that results in the longer time frame and hence disaster reduction. However, in the real world a number of risk outcome scenarios are possible. Uncertainty might be reduced through advances in risk assessment, but commitment to bringing about change can remain unmatched personal and wider political commitment. For example, we have the evidence of climate change impacts on weather systems, but choose not to act on it. We know the link between poverty and disaster events, but fail to make the world more just. This locks us into a situation of high disaster impact under conditions of certainty, delaying any shift to more stable attempts to deal with threats, pending a moral and political change.

Box 6.2

Late lessons from early warnings

Harremoës *et al*. (2002: 218) provide twelve lessons based on a series of twentieth century case studies of environmental and health crises as follows;

1. Acknowledge and respond to ignorance, as well as uncertainty and risk, in technology appraisal and public policy-making.
2. Provide adequate long-term environmental and health monitoring and research into early warnings.
3. Identify and work to reduce 'blindspots' and gaps in scientific knowledge.
4. Identify and reduce interdisciplinary obstacles to learning.
5. Ensure that real world conditions are adequately accounted for in regulatory appraisal.
6. Systematically scrutinise the claimed justifications and benefits alongside the potential risks.
7. Evaluate a range of alternative options for meeting needs alongside the option under appraisal, and promote more robust, diverse and adaptable technologies so as to minimise the costs of surprises and maximise the benefits of innovation.
8. Ensure use of 'lay' and local knowledge, as well as relevant specialist expertise, in the appraisal.
9. Take full account of the assumptions and values of different social groups.
10. Maintain regulatory independence from interested parties while retaining an inclusive approach to information and opinion gathering.
11. Identify and reduce institutional obstacles to learning and action.
12. Avoid 'paralysis by analysis', by acting to reduce potential harm when there are reasonable grounds for concern.

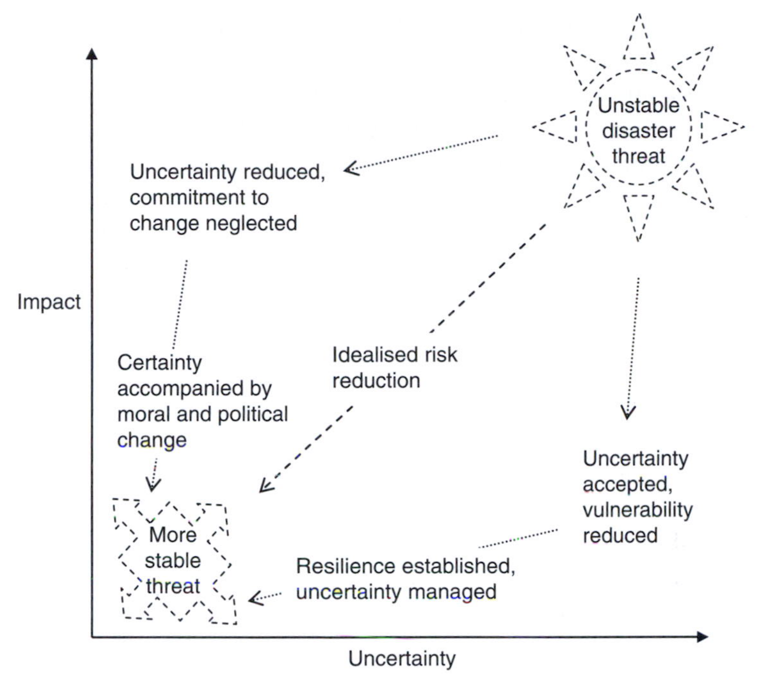

Figure 6.2 The relationship between disaster impact, uncertainty and risk reduction

An alternative route is where the knowledge base is inadequate, with consequent loss of effectiveness in the delivery of disaster risk reduction. However, reduced vulnerability, such as poverty reduction or making environments secure, progresses both hidden as well as known benefits of risk reduction substantially. There is a delay to disaster risk reduction in this instance through lack of knowledge. However, the second phase of this route leads to the possibility of establishing human resilience to the extent that uncertainty is increasingly managed and consequently reduced. This simple representation demonstrates disruption of disaster risk reduction through lack of political will and moral imperative, and through lack of commitment to knowledge creation through research, technology and awareness. Progress on each of these fronts represents a more direct and rapid shift to disaster risk stability.

Assessment of processes of change in development is in itself a form of risk assessment. This becomes particularly poignant in the context of poverty assessment. Perpetually dealing with risk and uncertainty is a central part of how people develop their own life chances and survival. The process of working with risk in emergency preparedness and disaster avoidance is both through risk assessment and the application of risk management. There are complex arrays of influences on our perception of risk. The role of different understandings of risks and of different levels of risk tolerance is crucial to emergency preparedness, disaster management and sustainable development. For example, together with various more quantifiable risk factors, notions of risk have an influence on the following:

- prioritisation in poverty/vulnerability reduction strategies;
- levels of preparedness in response systems;
- the state of the insurance industry.

Entire industries are engaged in the art of prediction. Their aims may be as diverse as the search for business continuity and cost-effectiveness (through market risk analysis), health care (i.e. targeting medicines and vaccine) and profit (as in the insurance business).

Varied motivations are revealed when probing further into what we refer to as the field of risk governance. Figure 6.3 presents a notional representation of risk governance factors that influence the cycle of decision making in risk assessment and management. It indicates that risk assessment is dependent upon the measurement of risks, knowledge and risk perception. Translating risk assessment into risk management

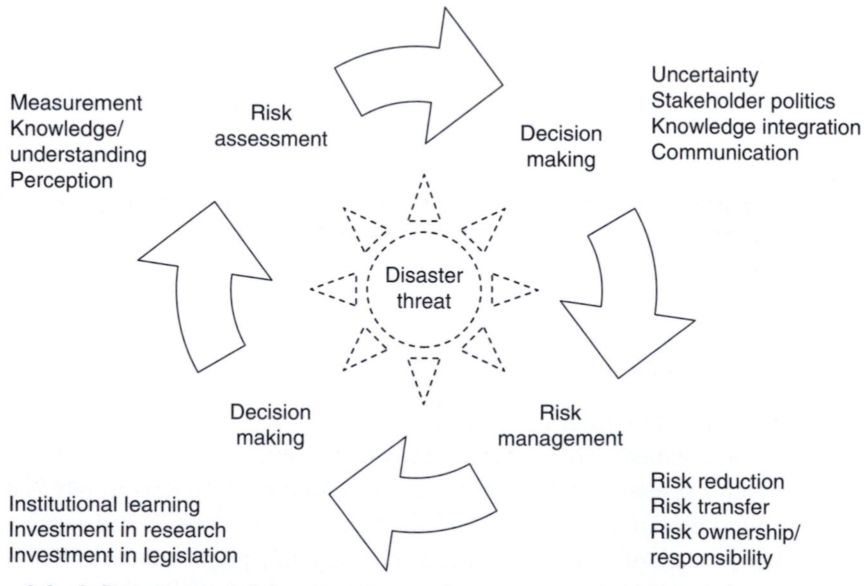

Figure 6.3 Influences on a risk governance cycle

decision making depends on the remaining uncertainty about the risks, the interests of different stakeholders in this process, the ability to wisely integrate different knowledge and to communicate them. Risk management is the implementation of the activities that have been judged by the risk assessment to be dependent on effectiveness in reducing risk, risk transfer, such as through insurance, and realisation of ownership and responsibility of risks. For example, if climate risks can be associated with processes that countries or companies own, such as manufacturing, ownership of that process translates into responsibility for the risks created. The adequacy of decision making in this cycle of risk governance depends on the extent to which risk management is learnt, ongoing research supported and the legislation developed.

A further aspect of risk governance decision making is to consider the circumstances within which we should adopt higher risk strategies. How do we decide what risks are unacceptable and who is it that actually decides? Something we do know is that decision making in uncertain situations is dependent on people's attitude to risk. The point of achieving better risk assessment is to arrive at better decision making to increase the capacity to risk assess and manage, and to provide insight for appropriate policy making. A sense of the nature of risk and how we can understand it is accentuated the closer we are to emergencies. Some of the core influences on more personalised notions of risk are summarised in Table 6.1.

Table 6.1 ***Some core influences on reactions to risk***

Core theme	*Core contextual influences*	*Intervening influences**	*Risk management implications*
Knowledge	Experience - age, activity, education	experience, lack of experience, education	Knowledge base for reacting to risk
Power	Structure - economy, politics, society	uneven development, technology, access and rights	Presence or not of constraints to being able to react to risk
Culture	Social origin - agency, tradition, faith	personality, altruism, strength of kinship, gender based risks, beliefs in immortality, faith and non-faith based traditions,	Rationale to choice, motivation or belief in reactions to risks
Environment	Environment - systemic and chaotic hazards, changing nature of places	vulnerable and resilient people in hazardous or safe locations	Place oriented interactions with risk

* Stakeholder reactions to the composition and context of risk

Some of the dilemmas of working with risk in disaster and development

There is no suggestion here that working with risk is a simple route to disaster reduction. The following are a few of the challenges that arise in working with risk:

1. In gaining knowledge and wisdom about a risk, we may uncover a lack of understanding about new risks and neglect knowledge and wisdom regarding old risks.
2. In reducing risk we may encourage further risks, whilst present increased risks may either offset or intensify future risks.
3. In not recognising a risk we may allow further risks to develop, and in recognising a risk we may construct new ideas about risks.
4. By avoiding hazards and vulnerability we can reduce a risk and by intervening in hazards and vulnerability we may manage risk for better or for worse.

Working with risk therefore requires that we understand the constituent components of risk, risk stakeholders and the development of risk outcomes. The right to take risks and to decide on what is a risk is likely to be based on mixtures of objective and subjective knowledge and understanding of risk unique in space and time.

Conclusion: managing risk as part of development

Early warning, prediction, uncertainty and risk are interconnected concepts underlying disaster avoidance policy and practice. It is normally not possible to identify accurately the time and place of forthcoming events. Disaster reduction and development will be advanced by greater commitment together with and strengthening of the knowledge base. Improving the knowledge base involves having the maturity to be able to live with uncertainty in a manner that builds preparedness and resilience to disaster events. Ultimately, disaster risk reduction is also a function of good governance, in that the responsibility for risk management always rests somewhere.

Discussion questions

1. What might be the most appropriate early warning approaches associated with different hazard and vulnerability interpretations of disaster? Consider the case of famines, disease epidemics, conflict and hurricanes/cyclones.
2. Explain how indicators can be used in an integrated approach to early warning.
3. To what extent do you think it is possible to predict major crises events?
4. Discuss the balance between knowledge and political will in disaster risk reduction in relation to the precautionary principle.
5. What is meant by risk governance?

Further reading

Kasperson, J.X. and Kasperson, R.E. (eds) (2001) *Global Environmental Risk*, London: Earthscan.

Linnerooth-Bayer, J., Löfstedt, R.E. and Sjöstedt, G. (eds) (2001) *Transboundary Risk Management*, London: Earthscan.

Strategy Unit (2002) *Risk: improving government's capability to handle risk and uncertainty*, London: UK Cabinet Office.

Tudor Rose (2005) *Know Risk,* www.tudor-rose.co.uk: Tudor Rose.

Tudor Rose (2006) *Real Risk*, www.tudor-rose.co.uk: Tudor Rose.

Tudor Rose (2008) *Risk Wise*, www.tudor-rose.co.uk: Tudor Rose.

UNISDR (2004) *Living with Risk: a global review of disaster reduction initiatives*, Geneva: (UNISDR).

Zschau, J. and Küppers, A.N. (eds) (2003) *Early Warning Systems for Natural Disaster Reduction,* Berlin: Springer.

Useful websites

www.odihpn.org: Humanitarian Practice Network (HPN), particularly the journal *Humanitarian Exchange.*

www.sra.org/glossary.htm: Society for Risk Analysis.

www.unisdr.un.org: the document 'Living with Risk' can be sourced at this site. The International Strategy for Disaster Reduction sponsored a series of Early Warning Conferences that can be accessed through their site. Also at www.ewc2.org.

www.fews.net: United States Agency for International Development (USAID) Famine Early Warning System.

www2.reliefweb.int: ReliefWeb provides regular updates on disasters and disaster threats from around the world, including links to key information for macro-scale monitoring and early warning.

www.irgc.org: International Risk Governance Council. See, in particular, at this site 'White Paper on Risk Governance: towards an integrative approach'.

Disaster mitigation, response and recovery

Summary

- Disaster mitigation reduces the impact of rapid or slow onset environmental, social and economic crisis through protection, resilience and adaptation strategies.
- Physical planning, including building design and engineering regulations, is an aspect of disaster mitigation and sustainable disaster recovery.
- Once people are caught in the wrong place at the wrong time without adequate forms of protection, hazards and vulnerability mitigation become emergency response and relief.
- Integrated emergency response variously depends on technology, appropriate organisation and human sensitivity to different development contexts.
- Humanitarianism provides a rationale for emergency relief with varied origins, interpretations and practical implications.
- Common standards in humanitarian assistance are to encourage an adequate service and accountability to intended beneficiaries and donors.
- Disaster recovery is a process of personal and institutional restoration and development. It is dependent on the varying way people are able to return to what they consider to be normality.

Introduction

This chapter turns to various aspects of responding to disasters in a developmental way. This is in the interest of the short-term relief and long-term recovery processes.

Disaster mitigation is about reducing the impact of events that cause human crises. This may be in ways as diverse as protecting people from environmental hazards through a defensive infrastructure, from abuse through policing and social services, and from financial ruin through emergency financial buffers. Another way of looking at mitigation is through a resilience approach, whereby the more environmental, social and economic assets people have, then the less they will feel the impact. In recent times, and with respect to climate change impacts in particular, the idea of adaptation has also come to the fore. The way in which environmental, social and economic processes adapt is embedded as a way of mitigating the impact of adverse climate changes (Adger *et al.* 2006; UNDP 2007). There are conceptual overlaps between some of the different disaster management related terminologies (Thywissen 2006). Care is needed in trying to separate out overly definitive definitions in the terminology of mitigation.

Whether through notions of protection, resilience or adaptation, mitigation is best thought of as ongoing rather than a one-off stage in a cycle of disaster activity. It has application to both rapid onset and slow onset disasters and is part of building defences, emergency response and recovery processes. This is because even rapid onset disaster events do not pass in a moment; rather, ongoing impacts of an event unfold over the minutes, hours, days, months and years following. This is one interpretation of what we consider to be repetitive or multiple disaster impacts. Referring to the example of the Pakistan Earthquake of 2005, the earth tremor was the first in a series of impacts. It caused entire town- and village-scapes to collapse. The impact of the tragedy for survivors was the discovery of a missing relative dead in the rubble hours or even days later. There was further suffering in the risk of aftershocks, onset of desperation, increased isolation through loss of the roads, breakdown of food supply, the onset of an early winter, long-term psychological impacts and disaster relief dependency. The reconstruction process includes mitigation measures in the form of rebuilding using earthquake resistant designs, people's rehabilitation physically and mentally, and restoration of the capacity and the will to protect against further tragedy. Where disaster impacts have been effectively reduced through investment in mitigation, response and recovery operations have been given a comparatively better chance of success. It costs less to mitigate than to carry out response, reconstruction and long-term recovery. However, one of the limitations of emergency response systems is that they are designed for the shorter term, relief operations being reactive and constrained by resources. With increasing numbers of disasters occurring and now known about, there is a concern that not enough relief aid will be available, and that shifting budgets towards relief detracts from development.

Recovery is often a longer-term process lasting months and years and consequently dependent on the process of sustainable development. The protection, resilience and adaptive capacity needed to restore people's sense of safety are factors that influence response and long-term recovery. Responses to disaster are in effect going on over a very long time-frame if they are addressing the underlying influences on disaster events. Short-term responses that do not address underlying problems risk putting out fires only to leave others nearby to smoulder until tragedy returns anew. For example, a humanitarian relief operation can be effective at getting consignments of grain in place, but will have achieved relatively little if a longer-term food security initiative is not established. For many years, donors, NGOs and other governmental departments have sought to find ways of adapting relief to development activities post-crises. This was one of the key contexts of an earlier text on *Disaster and Development* by Fred Cuny during the 1980s (Cuny 1983). Whilst the transition from relief to development is clearly what sustainable recovery would suggest is required, institutions also adopt this for logistical reasons, even their own 'business continuity'. It can be practical once relief organisations are in place following a major disaster event for them to become involved in the development process of the area within which they find themselves. An issue in this respect are, however, the extent to which this prevents a genuinely sustainable recovery process.

The resiliency, adaptive capacity, safety and fragility of disaster response are dependent on issues of local ownership, inputs of knowledge, control and cooperation across multiple stakeholders. The UK has a system of category 1 and 2 responders and of gold, silver and bronze command, the remit of each of these categories being pre-designated. This is described in more detail later in the chapter. The implementation of the Civil Contingencies Act in the UK in 2004 was accompanied by greater realisation of the importance of integrated emergency management. The emergencies services are integrated because of recognition of the need for many key players, roles and responsibilities to be coordinated during a major incident. Improved information communication technology (ICT) has made its contribution, and many aspects of emergency operations require that this is used as a simplifying process. For example, emergency responders from various parts of the police, fire, ambulance services and beyond expect to have access to commonly calibrated and transferable procedures and equipment. Emergency responders are frequently concerned about being on the same wavelength, which in recent years has created a demand for improved and integrated communications technology. Beyond the technology, emergency response in a globalised world in which mass emergencies take place means that responders from different contexts find themselves working together in other cultural contexts. This requires sensitivity to varied ways of operating, including, for example, how the dead may be dealt with (Scanlon 1998) and the often neglected specifics of gender in disasters (Fordham 2003; IFRC 2006).

Identifying standards in disaster response has much to do with what we mean by humanitarian perspectives. Varied trends exist in the literature. These include humanitarianism as a cultural imperative, in which rights and minimum standards can be applied based on a Humanitarian Charter approach (Sphere project 2004). Another section of the literature looks at humanitarianism with a strong critique, identifying it more as a supply driven industry subject to global political economy (Middleton and O'Keefe 1998; Anderson 1999; Smillie and Minear 2004). Humanitarian assistance, once embedded in a local culture, might also be considered the antithesis to a disaster and development approach should local coping and strategies be undermined and neglected in the process. Such concerns have gradually become better highlighted in evaluation of humanitarian assistance (ALNAP 2002), but awareness and knowledge can take many years to catch up with practice.

The Hyogo Framework of 2005, referred to earlier in this book (see Box 1.3, p. 43), has put a resilience approach to disaster risk reduction firmly in the minds of the international community of scholars and policy makers. Moreover, most would understand this to extend to building disaster resilient people, communities and institutions not only in a preventative mode, but also in response systems. Whilst terminologies move up and down the ranks in terms of their contemporary usage, for emergency services the challenge of putting resilience into response is not new. Robust working systems, training and personal integrity of personnel are all qualities expected in emergency responses worldwide and these can be aspects of resilience. Putting resilience into response and recovery has wider resonance. It means to respond and recover with adequate forms of protection, rights, capacity to adapt, and the recreation of a sense of safety and wellbeing. The manner in which this can happen depends on the progress of integrity in environmental, social and economic developments.

Reducing the impact of disasters

McEntire (2007) provides a convincing summary of how the recovery phase of a disaster is an ideal time to be designing mitigation strategies. Interest in the wake of a disaster is much greater than usual. Whilst awareness is raised, a 'mitigation action plan' can be established. This identifies actions against vulnerabilities. It is a strategy that in the first instance might identify mitigation actions such as whether or not to relocate people to a different area or rebuild. However, even where there has been catastrophic destruction of life, livelihood and property, there is often a tendency for people to want to rebuild in the same place. Mileti and Passarini (1996, cited in McEntire 2007) identify reasons for wanting to build in the same area as destruction occurred as being due to a mix of political, cultural, economic and psychological factors. This has been observed in the wake of the floods in Mozambique. (Collins and Lucas 2002) (Box 7.1).

Alexander (2000) balances risk mitigation actions with those of risk amplification. Furthermore, a mitigation action plan can also focus on resilience factors. The approach of working with resilience rather than just against vulnerability, suggested at several points in this book, also fits the model. An ideal strategy would therefore be one that mitigates hazard, vulnerability and risk and amplifies protection, resilience, adaptation and safety.

Developing disaster resistant infrastructures and communities

Ozerdem reminds us that 'mitigation strategies can be classified as structural, including the construction of hazard resistant buildings, windbreaks or dams; or non-structural, including building codes, land use policies, and procedures for forecasting and warning' (2003: 210) Disaster resistance in this sense depends on what technological and infrastructural investments get made alongside policy shifts that together mitigate hazards and vulnerability. Pelling (2003b) plays mitigation off against adaptation in pointing out that the cost of mitigation can be afforded by wealthy countries, whilst policy shift towards greater adaptation shifts responsibility and cost to developing nations. This would

Box 7.1

Why do people move back home rather than mitigate risk by going somewhere safer?

After the flood waters had subsided, people at Xai-Xai in southern Mozambique suggested that factors other than a sense of elevated fear of floods determined whether or not people were returning to their home area. One factor cited by residents of Ndambene 2000 (a resettlement camp near to Xai-Xai) was that they wanted their children to stay near to a school and these are provided close to or inside the areas to which they have been displaced. Often families returned in part, insomuch as some members went back to cultivate but left the children in the proximity of the emergency resettlement area schools. The relatively fertile soils for cultivation and an ongoing sense of ownership in the home areas determined a strong economic pull back to the place from which people were displaced. These areas include the zones nearest the river, which are the most fertile and easy to irrigate, but also at most risk from further flooding. The higher safer areas can be problematic to irrigate, as people do not own pumps. A further return pull factor for those displaced to sites further away was that they preferred to reside near to where their ancestors are buried. The greatest pull back to the scene of disaster, which in this instance was the area of greatest future risk, was first and foremost to do with recreating livelihoods. Individuals weighed up the risks of further flooding on the low-lying areas with the risk of not being able to recreate a livelihood in the resettlement areas. Mitigation of future crises in this sense was ignored.

Despite expressing appreciation for the efforts that had been made by the emergency services, a resident interviewed at Ndambene 2000 and health staff pointed out that resettlement areas were not secure socially. In the emergency centres and resettlement areas people generally found themselves housed in new communities of strangers, rather than alongside their former community. Whilst there was relative physical health security through implementation of health care programmes in these areas, residents were concerned about social insecurity through a breakdown of community cohesion. In the areas that were destroyed there had been a strong tradition of helping your neighbour and not stealing from them. This allegedly had broken down in the resettlement areas, where people ended up living with strangers. However, an old woman pointed out that, despite these difficulties, the trauma she had experienced with the floods and the death of some of her family in her home area made her determined not to return there. A visit to Bairro Ximanga in the Limpopo Bridge high flood zone indicated that some had clearly opted to return. One of the interviewees explained that people returned to this area because of economic and cultural ties to that particular stretch of land. One woman said that the respondent who said she had chosen to remain in the resettlement area because of trauma was lying, and that the death of family members in a home area was no reason to stay away. She also went on to emphasise that one problem in the original places of habitation on the flood plain was that there were no primary schools nearby. Children who went to school had to travel to areas of higher elevation some distance away, tried to get into schools in the town or post-disaster were making use of the emergency resettlement area schools. From these interviews, it could be appreciated that there were a variety of ways in which people gambled with different types of mitigation possibilities, determining the location within which they would reside.

On the subject of implementing a mitigation strategy for next time around, the 71-year-old Secretary of the Bairro Ximanga responded that the flood had been of 'biblical proportions' and therefore was a 'one-off' that would not come back in 500 years. Clearly many people have taken this view since most of the low areas in Xai-Xai town and its surroundings have been resettled, particularly the poorest Bairros (high density population sub-areas of the city), where the attraction had been to be close to the town at any cost. Some people residing in those areas had a longstanding tie to them dating back to the time when they had been war refugees displaced to these urban locations for safety. People who have known more suffering are likely to gamble with a higher order of risk and disregard mitigation measures to preserve a measure of livelihood security. A resident interviewed at Bairro Three in Chókwè indicated that in his area people tended to resettle the flood plain as their permanent abode rather than as one of two residences. It is more difficult to escape flooding catastrophes in these locations as the safe areas are many kilometres away. However, land tenure and land productivity are major pull factors and something that could not be obtained in resettlement areas, where housing is very close together and where there had generally not been allocations of secure land tenure for cultivation. Some people chose to negotiate a set of hazards by migrating between two areas of 'higher' and 'lower' risk. However, by way of contrast with Mozambique, the people of New Orleans were banned from moving back to their former residences as a way of mitigating secondary hazards.

Source: adapted from Collins and Lucas (2002).

appear to be the case when considering climate change and the difficulty that low-lying parts of the developing world will face in paying for sea defences to cope with sea level rises. In most instances it will prove too vast an undertaking, leaving populations with the only other option of adapting their livelihoods and survival strategies at great personal social and economic cost. However, a number of strategies have been launched that address infrastructural mitigation for key infrastructures in developing areas. These include a particular focus on protecting schools and hospitals. Information on both of these campaigns can be accessed via the UNISDR website provided at the end of this chapter (see p. 249). An example of typhoon resistant housing for the poorest of the poor in the Philippines is provided by Diacon (1997). More examples can be accessed through the literature of organisations such as PAHO, Action Aid and NSET (websites on p. 249).

Integrated emergency management as development

The combination of people in hazardous places, at hazardous moments and without adequate forms of protection means that early warning and preventative action have failed. There is at this point a dependence on short-term emergency management. Though self-coping comes into play, those caught up in the incident are usually reliant on others for assistance. The likelihood of survival is enhanced through adherence to pre-prepared emergency response procedures. This adherence is likely to vary in approach from country to country, but with some common features. One is the objective of creating order and clarity in response where there might otherwise be confusion and a deepening of the crisis. Emergency responders are also subject to the impact of the events. Coping amongst field workers in emergency settings is a known issue for humanitarian organisations and other emergency responders but surprisingly little attention has been paid to this. Paton and Auld identify this as an important aspect of resilience, in this context being the 'capacity to make choices regarding the allocation of resources . . . and performing tasks under highly stressful circumstances' (2006: 267).

In the UK, the implementation of the Civil Contingencies Act (CCA) in 2004 (Box 7.2) confirmed an integrated strategy as one of the priority areas for development. The approach had its origins in the Home Office disaster response oriented publication of 1994 on *Dealing with Disasters,* as follows:

> Under the principles of integrated emergency management, the response to an emergency should concentrate on the effects rather than the cause of the disaster and, wherever possible, should be planned and undertaken as an extension of normal day to day activities.
>
> (Home Office 1994: 3)

However, according to Resilience UK, a web resource set up to further the aims of the CCA for the Cabinet Office in 2004, Integrated Emergency Management (IEM) comprises six related activities: anticipation, assessment, prevention, preparation, response and recovery. The quotation is making a point about post-emergency elements of IEM.

Box 7.2

The UK Civil Contingencies Act

This came about following realisation that although a developed country the UK was not prepared for major incidents. The evidence had been the challenges encountered in managing the foot and mouth epidemic amongst livestock, fuel crisis and localised flooding of the late 1990s and early 2000s. However, in many respects the development of the Act ended up being driven by the global situation. As with the changes that occurred in the Federal Emergency Management Agency (FEMA) following the destruction of the World Trade Center in the United States on 11 September 2001, the agenda became guided by the needs of homeland security.

The Act is separated into two substantive parts: local arrangements for civil protection (Part 1) and emergency powers (Part 2). The statutory guidance which supports Part 1 of the Act allows for the making of temporary special legislation (emergency regulations) to help deal with the most serious of emergencies.

The aim of the CCA is to deliver a single framework for civil protection in the UK to:

- ensure consistency of activity across tiers of government;
- set out clear responsibilities for front-line responders at the local level, to ensure that they can deal with the full range of emergencies from localised major incidents through to catastrophic emergencies;
- modernise the legislative tools available to government to deal with the most serious emergencies, providing for greater flexibility, proportionality, deployability and robustness;
- link with the practical civil protection measures which the government has already put in place to build capabilities.

To implement the strategy, institutional players in emergency response have been identified as follows:

Category 1 responders

Local authorities
- All principal local authorities

Government agencies
- Environment Agency
- Scottish Environment Protection Agency
- Maritime and Coastguard Agency

Emergency services
- Police Forces
- British Transport Police
- Police Service of Northern Ireland
- Fire Authorities
- Ambulance Services

NHS bodies
- Primary Care Trusts
- Health Protection Agency
- NHS Acute Trusts (Hospitals)
- Foundation Trusts
- Local Health Boards (Wales)
- Any Welsh NHS Trust which provides public health services
- Health Boards (in Scotland)
- *support Health Authorities*

Category 2 responders

Utilities
- Electricity
- Gas
- Water and Sewerage
- Public communications providers (landlines and mobiles)

Transport
- Network Rail
- Train Operating Companies (Passenger and Freight)
- Transport for London
- London Underground
- Airports
- Harbours and Ports
- Highways Agency

Government
- Health and Safety Executive

Health
- The Common Services Agency (in Scotland)
- Scottish Health Authorities (SHAs)

Category 2 responders communicate and cooperate under the category 1 responders.

The statutory duties of these institutions are as follows:

Category 1:

- risk assessment
- emergency planning
- business continuity planning
- warning and informing
- cooperation
- information sharing

Category 2:

- cooperation and information sharing

Local authorities only:

- promoting Business Continuity Management (small and medium-sized enterprises)

The government has set up a system to implement the agenda through Regional Resilience Forums (RRF). These in turn contain Local Resilience Forums (LRF) and Working Groups sub-groups for:

- Risk Assessment
- Training & Exercise
- Business Continuity
- Warning & Informing

There are also 'task and finish' groups for specific risks such as flooding.

A system of command and control is used to implement three levels of responses to major incidents, as follows:

- Operational Response (Bronze Command) – with the police having overall control:
  - The emergency services response to any incident (emergency)
  - Rapid Response
  - Front Line Ambulance Response
  - Basics, i.e. doctors
- Tactical Response (Silver Command):
  - Emergency Services & Local Authority
- Strategic Response (Gold Command):
  - Strategic Coordinating Group, Scientific and Technical Advice Cell (STAC), Recovery Working Group

Some of the more detailed roles and responsibilities of selected category 1 responders beyond the primary goal of saving lives are as follows:

- Police
  - coordination of the emergency services and other agencies;
  - controlling access to and regress from the site;
  - the protection and preservation of the scene;
  - evacuation procedures, in consultation with the other emergency services and the local authority;
  - investigation of the incident in conjunction with other investigative bodies;
  - collation and dissemination of casualty information;
  - identification of victims on behalf of the coroner;
  - restoration of normality at the earliest opportunity.
- Fire Service:
  - firefighting;
  - rescue in the event of persons being trapped in a fire, wreckage or debris;
  - assist in essential and appropriate salvage operations;
  - dealing with released chemicals or other hazardous materials in order to render the incident site safe;
  - responsible for the health and safety of all emergency service and other personnel working within the inner cordon (except for actual/suspected terrorist incidents).
- Local Authorities:
  - provide support for the emergency services;
  - continue normal support and care for the local and wider community;
  - use resources to mitigate the effects of the emergency;

- coordinate the response by organisations other than the emergency services, e.g. voluntary organisations, clergy, etc.;
- facilitate the rehabilitation of the community and the restoration of the environment;
- Chair Recovery Working Group.

- Ambulance Service:
  - provision of paramedic services;
  - establishing a casualty clearing point and ambulance loading point;
  - MERIT Teams and Medical Incident Commander;
  - decontamination;
  - responsible for naming receiving hospitals to which casualties will be taken;
  - transportation and continuing treatment of casualties en route to hospital;
  - provide a focal point for all NHS/medical resources;
  - HART Teams – piloted.
- Acute Hospitals:
  - reception of casualties in Accident and Emergency (A&E);
  - back-up facilities in the rest of the hospital;
  - MERIT/Medical Incident Officer (unless other local arrangements are agreed);
  - decontamination facilities;
  - counselling and support in partnership with other agencies.
- Primary Care Trusts:
  - duty to protect and promote the health of the public;
  - initiating and supporting the public health response;
  - delivery of primary and community health services;
  - mobilisation of community resources;
  - support of NHS infrastructure.
- Health Protection Agency:
  - protect the community against infectious disease and other dangers to health;
  - expert advice to the Department of Health on health protection policies and programmes;
  - specialist support to the NHS and other agencies;
  - specialist emergency planning advice to the NHS;
  - resources to support Specialised Commissioning Groups (SCGs) and Retail Competition and Consumer Choice (RCCCs) – activation of STAC;
  - training and exercise support on behalf of the DH.

- Strategic Health Authority:
  - strategic command and control of widespread major incidents;
  - performance management of NHS organizations.
- Department of Health Emergency Planning Unit:
  - develops, issues and maintains national guidance/policy;
  - contributes to central government response, e.g. Cabinet Office Briefing Room (COBR) or the Civil Contingencies Committee (CCC);
  - oversees and supports NHS response;
  - national coordination of the NHS during a complex national emergency incident through the Major Incident Coordination Centre (MICC).

Source: Christine Mathieson, Deputy Regional Health Emergency Planning Adviser, Health Protection Agency North East, citing the Civil Contingency Act (CCA) of 2004 in 2008.

Humanitarianism and accountability

This section looks at accountability in the instigation of aid. There have been frequent accusations not only that the nature of a crisis can be misinterpreted, but that the response can be badly informed, inappropriate and can even make a crisis situation worse. However, there is rarely unanimous agreement about these accusations. Practitioners are wary of what appear sometimes to be 'armchair commentators'. The accountability debate can, however, be usefully focused on what was justified, what was successful and what was not, so that lessons are learnt and improvements are made. Whilst there are differences of opinion as to how beneficial 'intervention' and 'non-intervention' in mass disasters can be, arguably to do nothing (either through aid or advocacy) is to go backwards. These are moral and ethical issues regarding aid intervention. The intention in this book has been to avoid being overly polemical on emotive and complex topics in the interests of drawing attention to a range of arguments. Learning the lessons from extreme disasters, such as famines in Ethiopia, has raised issues of crisis definition, accountability, the politics of aid, timing, not learning lessons, communication, appropriate aid content, and rights, to name a few.

As an introductory guide to the classification of humanitarianism and minimum standards that might be expected in the world of the humanitarian response agencies, the Sphere Project is an interesting case study. It was launched in 1997 and reviewed and re-released as a code of conduct for the range of NGOs representing within it in 2004. It in part stems from concerned individuals within the aid system who consider that humanitarianism has a collective responsibility to improve the effectiveness and accountability of humanitarian response. The Sphere Project has been considered to represent a Humanitarian Charter. This in turn is based on:

- the Universal Declaration of Human Rights;
- Geneva Conventions and Protocols;
- Refugee Law.

The rationale of Sphere is clear in the following statement from its main document:

> We reaffirm our belief in the humanitarian imperative and its primacy. By this we mean the belief that all possible steps should be taken to prevent or alleviate human suffering arising out of conflict or calamity, and that civilians so affected have a right to protection and assistance.
> (Sphere Project 2004: 16)

The following items of legal and institutional development map the more detailed progression of events considered to currently inform the *Humanitarian Charter and Minimum Standards in Disaster Response:*

- Universal Declaration of Human Rights, 1948;
- International Covenant on Civil and Political Rights, 1966;
- International Covenant on Economic, Social and Cultural Rights, 1966;
- International Convention on the Elimination of All Forms of Racial Discrimination, 1969;
- the four Geneva Conventions of 1949 and their two Additional Protocols of 1977;
- Convention relating to the Status of Refugees 1951 and the Protocol relating to the Status of Refugees, 1967;
- Convention against Torture and Other Cruel, Inhuman or Degrading Treatment or Punishment, 1984;
- Convention on the Prevention and Punishment of the Crime of Genocide, 1948;
- Convention on the Rights of the Child, 1989;
- Convention on the Elimination of All Forms of Discrimination Against Women, 1979;
- Convention relating to the Status of Stateless Persons, 1960;
- Guiding Principles on Internal Displacement, 1998;

(Sphere Project 2004: 313)

Three core principles flagged by the Sphere Project have been:

1 the right to life with dignity;
2 the distinction between combatants and non-combatants;
3 the principle of non-refoulement.

The right to life with dignity means respecting cultures, gender and different ways of doing things, as well as making sure people have what they need to stay alive. Following the confusion surrounding the Rwandan crisis as to how to support a mix of civilians and combatants, there is acknowledgement of the need to distinguish between the two. 'Non-refoulement' is the principle of not sending those who have undergone and fled persecution back to the area where this occurred. Emphasis is put on the responsibility of governments to respond to humanitarian disasters and, in particular, of governments to assist all people within their borders. However the Sphere handbook itself is meant as a simplified reference handbook for aid workers, and is available in multiple languages.

Minimum standards in disaster response (Sphere guidelines)

Sphere Project minimum standards are set for the following:

- water supply, sanitation and hygiene promotion;
- food security, nutrition and food aid;
- shelter, settlement and non-food items;
- health services;
- standards common to all sectors.

Minimum standards are there to help agencies plan and manage in emergencies. There is a set of standards that are regarded as common to all sectors (Box 7.3) and more detailed ones, both of which can be monitored by use of key indicators. For example, in the case of water supply it is recommended that at least 15 litres of water per day per person is an indicator of a minimum standard or 'right'. Though it is complicated to quantify or even qualify with indicators what it means to live with dignity, the approach represented by Sphere is nonetheless a step towards accountability awareness.

Box 7.3

Sphere standards presented as common to all sectors:

Common standard 1: *participation*

The disaster affected population actively participates in the assessment, design, implementation, monitoring and evaluation of the assistance programme.

Common standard 2: *initial assessment*

Assessments provide an understanding of the disaster situation and a clear analysis of threats to life, dignity, health and livelihoods to determine, in consultation with the relevant authorities, whether an external response is required and, if so, the nature of the response.

Common standard 3: *response*

A humanitarian response is required in situations where the relevant authorities are unable and/or unwilling to respond to the protection and assistance needs of the population on the territory over which they have control, and when assessment and analysis indicate that these needs are unmet.

Common standard 4: *targeting*

Humanitarian assistance or services are provided equitably and impartially, based on the vulnerability and needs of individuals or groups affected by disaster.

Common standard 5: *monitoring*

The effectiveness of the programme in responding to problems is identified and changes in the broader context are continually monitored, with a view to improving the programme, or to phasing it out as required.

Common standard 6: *evaluation*

There is a systematic and impartial examination of humanitarian action, intended to draw lessons to improve practice and policy and to enhance accountability.

Common standard 7: *aid worker competencies and responsibilities*

Aid workers possess appropriate qualifications, attitudes and experience to plan and effectively implement appropriate programmes.

Common standard 8: *supervision management and support of personnel*

Aid workers receive supervision and support to ensure effective implementation of the humanitarian assistance programme.

Source: Sphere Project (2004: 313).

Sphere has also adopted the Principles of Conduct for the International Red Cross and Red Crescent Movement and NGOs in Disaster Response Programmes, as follows:

- The humanitarian imperative comes first.
- Aid is given regardless of the race, creed or nationality of the recipients and without adverse distinction of any kind. Aid priorities are calculated on the basis of need alone.
- Aid will not be used to further a particular political or religious standpoint.
- We shall endeavour not to act as instruments of government foreign policy.
- We shall respect culture and custom.
- We shall attempt to build disaster response on local capacities.
- Ways shall be found to involve programme beneficiaries in the management of relief aid.
- Relief aid must strive to reduce future vulnerabilities to disaster as well as meeting basic needs.
- We hold ourselves accountable to both those we seek to assist and those from whom we accept resources.
- In our information, publicity and advertising activities, we shall recognise disaster victims as dignified humans, not hopeless objects.

The Red Cross and Red Crescent Societies' humanitarian principles are based on the themes of humanity, impartiality, neutrality, independence and 'voluntariness'.

It is important to remember that *minimum standards* have been demanded in the wake of harsh lessons learnt. There has been pressure to coordinate and standardise better in the context of the multiple interest humanitarian machinery that must operate in complex situations, though varying interpretations of 'standards' suggests there will continue to be some variation in practice. Agencies have tended to have individual strategies, and the Sphere Project represents an attempt to draw these together. Some of the criticisms of this approach have been that it represents a tenuous link between the rights of affected populations and technical standards of interventions. These aspects of humanitarianism are founded on a right to assistance but do not exist in international law. A further criticism is that its usefulness is in question in terms of trying to promote universal standards where for technical performance relative measures would be more appropriate. Humanitarian assistance can end up being based on technical assistance emanating from Northern agencies. In the context of emergency relief, learning based on the participation of affected populations in decision making is arguably more prone to being neglected than in the case of development programmes.

The nature of humanitarian intervention, particularly in complex emergencies, is a mix of moral, political, accountability and resourcing issues. Simply having goodwill has proven not to be sufficient in responding to humanitarian emergencies. Beyond emergency response in the short term lies the challenge of sustainable recovery. This is about getting development out of disaster and requires long-term commitment and minimum standards in development assistance.

Multilevelled recovery and the reconstruction balance sheet

There are no simple models that describe how recovery takes place. This is because disasters are unevenly distributed, investments in recovery range from zero to billions of currency, and because people's group and individual recovery capacity and motivation vary.

One idealised notion is that the experience of disaster creates an opportunity to build back stronger. There is evidence of this when in the wake of a major event people organise their quest for wellbeing anew. The catastrophes that Hurricanes Georges and Mitch imposed on the Caribbean and Central America, exposing vulnerability in their track, led to a range of politically influenced recovery reactions. Damaged countries were paying debt payments to the IMF and World Bank during the disaster recovery period, whilst Scandinavian countries and France were choosing to write them off (Mowforth 2000). Community based initiatives were not encouraged in some locations by national governments as there was a tendency to equate these with left-wing liberation movements. What those with power decide to do in the aftermath of a disaster is governed by many considerations other than humanitarianism, but some simple choices can be made. For example, following Cyclone Sidr in Bangladesh in 2007 many of the micro-credit initiatives that famously hold local people in debt in those areas conceded to pressure to write off the debts. This might be seen as a charitable act of solidarity by fellow country people, though it is also in reality a pragmatic step to take. People who have lost everything in a cyclone are unlikely to be able to make the payments in any case.

Whilst identification of strategies toward long-term recovery will depend on what type of crisis has occurred, a general principle is that it is a multilevelled and staged process with critical decision making along the way. Whilst emergency relief in the short term is likely to be highly centralised, the transition to recovery is the developmental process of people taking ownership of their own wellbeing. It is usually not possible to identify the point at which emergency relief stops and self-help and investment begin. This is because recovery involves a complex of localised factors that either stimulate the process or do not. The political and economic context of the location has proven to be crucial. In Sri Lanka and Indonesia there was clear evidence of a reduction of hostilities between groups at war with each other during and shortly after the crisis. However, in the longer term conflict returned in Sri Lanka to a crisis level as bad as before, foreign aid possibly having made the situation even more complicated. In looking at recovery, the wider picture of environmental change must also be considered. Globalisation and climate change are just two of the forces that have a bearing on whether there will be recovery or a perpetuation of the degradation that major emergencies bring. The range of development issues that determine the success of recovery were presented as a theme for the *World Disasters Report* of 2001 (IFRC 2001).

Whilst structurally determined recovery tends to make up the external inputs to a disaster situation, the other part of what happens next concerns social coping mechanisms, community, local government, local leaders and organisations, community resources, cooperation and communication. A local media service can play an important role here in communicating between survivors around the area. The community level recovery process and the structures that can be used to facilitate it are along the lines of what is required for the preparedness and resilience building aspects of community development. For example, the Risk and Resilience Committees referred to in Chapters 5 and 6 of this book, which help to manage risks in the community, can also be the basic local structure from which the basics of recovery might be coordinated. Village level response units of one sort on the other are common throughout much of the world. Linking risk assessment, preparedness, early warning, resilience building and post-disaster recovery procedures into a more seamless community level organisation respected and encouraged by the state and private sector is a greater challenge.

Family level recovery depends on savings and resources held elsewhere, the strength of family, friends and kinship networks prior to the disaster event and how much of these has survived. Coping is also influenced by previous experience with disaster and accumulated resilience in that respect. The dynamic between this level of recovery and of the community as a whole is a process of recreating security and livelihoods. Reconstruction programmes typically require a combination of institutional strengthening and a governance infrastructure for reducing the vulnerability of individuals. The vulnerable groups in reconstruction programmes might be determined in terms of class, caste, gender, age, political affiliation, religion, tribal group and so forth. The experience of development assistance, particularly post-conflict, is one where restructuring can favour those being restructured or those controlling the restructuring.

The balance sheet for a programme of reconstruction might variously reflect the following foci:

- Social sectors:
  - education – particularly primary education;
  - health – physical and mental;
  - culture – rehabilitation of culture institutions and infrastructure;
  - women's affairs and social welfare – includes family reunification.

Plate 7.1 Food aid being delivered along the Zambezi river

Source: Author.

- Other infrastructures
  - public works and housing – road repairs, water supply, sanitation facilities;
  - transport and communications, i.e. railways;
  - mineral resources and energy, i.e. rehabilitation of electricity supply, gas;
  - justice – facilitate the recreation of judicial system and infrastructures;
  - state administration – rehabilitation of town and district councils. office equipment and a means of transport.
- Economic sectors:
  - agriculture – seeds, tools, restocking livestock, veterinary vaccines and equipment;
  - industry and trade – funding/loans for selected elements of the private sector. Support for public sector industry dependent on governance context;
  - fisheries – repair fishing boats, distribute nets.

Plate 7.2 House reconstruction in post-tsunami Sri Lanka

Source: Janaka Jayawickrama, Northumbria University.

- Resettlement:
  - urban context – allocation of housing plots, building, modification to architectural designs;
  - rural context – land allocation and redistribution. Disbanding of emergency accommodation centres.
- Other:
  - training for public awareness programmes – participation of all stakeholder groups, with emphasis on civil societal representation;
  - establishing of civil protection teams;
  - revisit and adapt early warning systems;
  - establish contingency plans that are tested through running contingency exercises;
  - identify safe locations;
  - pre-position resources needed to cope with further onset of disaster;
  - improved hazard and risk mapping.

(based on Oxfam 1995)

Conclusion: building avoidance, mitigation and accountability into response and recovery

Disaster mitigation, response and recovery are long-term commitments. In the ideal world, recovery would be resilient and sustainable if the vulnerability experience post-disaster is reduced compared with prior to the disaster taking place. However, the extent to which disaster prevention and sustainable development principles are put into response and recovery is a matter of rights and good governance. This operates in contexts of environmental, social and economic development grievances or opportunities whereby improvements can only be made through their proper representation. Management of emergencies and relief is very varied between developing and post-development contexts such as the UK, which has a Civil Contingencies Act and disaster response and mass emergency relief operations associated with inputs into developing regions. Quality of organisation, coherent standards of delivery, accountability and long-term sustainability planning are crucial in each instance. Developing areas are often dependent on humanitarian relief during crisis, which in itself has a very limited life span. Inasmuch as disaster impacts unfold unevenly around the world and within localised cohorts of people, recovery has also proven to be dependent on structural and societal evolution and integrity.

Discussion questions

1. Identify some common approaches to disaster mitigation. To what extent can they be part of everyday life?
2. What are some of the practical ways in which resilience can be built into disaster response and recovery?
3. In what ways might there be similarities and differences in approach between the UK Civil Contingencies Act and humanitarian responses more widely?
4. Are minimum standards in humanitarian relief attainable?
5. In what way are there development alternatives to the types of emergency responses you have encountered in this chapter?
6. Why is post-disaster recovery a varied experience?

Further reading

Alexander, D. (2002) *Principles of Emergency Planning and Management*. Harpenden: Terra Publishing.

McEntire, D.A. (2007) *Disaster Response and Recovery*, Hoboken, NJ: Wiley.

Sphere Project (2004) *Humanitarian Charter and Minimum Standards in Disaster Response*, Geneva: Sphere Project.

Useful websites

www.adpc.net: Asian Disaster Preparedness Centre.

www.adrc.or.jp: Asian Disaster Reduction Centre.

www.unisdr.org: United Nations International Strategy for Disaster Reduction (UNISDR). Includes information and links to campaigns such as those for disaster resistant schools and hospitals.

www.paho.org: Pan American Health Organisation programme on protecting hospitals from disasters.

www.ukresilience.info: the Resilience UK website.

www.actionaid.org: Action Aid, an NGO with a programme to protect schools from disasters and provide disaster reduction education.

www.nset.org.np: Nepal Society for Earthquake Technology (NSET), an NGO that carries out campaigns to retrofit buildings and alter building planning regulations in Nepal to protect people against earthquakes.

www.sphereproject.org/: the Sphere Project.

www.odihpn.org: ODI Humanitarian Practice Network.

8 Conclusion

Summary

- **Many possibilities are available for disaster reduction and sustainable development, depending on investment and progress in science and technology, risk behaviour and political will.**
- **Although they emerged separately over the years, disaster reduction and sustainable development initiatives are essentially part of a common agenda.**
- **Sustainable environmental, social and economic futures are dependent on applying the right risk reduction decisions, at the right time, in the right place and with the right people.**
- **Progress in disaster reduction and sustainable development is qualified in terms of people's transition from vulnerability to wellbeing.**
- **Rights, responsibilities and governance are core aspects of the disaster and development nexus.**
- **People do more than just cope with disasters, they interact with them to build greater prosperity and a secure future, getting development out of disaster.**

Opportunity and synthesis in disaster and development thinking

The title of and basic idea behind this book are not new to the world of disaster and development scholars, but have demanded further interpretation and consolidation. Much of the content of this book points to opportunities for mitigating the impact of critical hazards through vulnerability reduction. Perspectives on the links between disaster and development have moved on from simply combining relief and development activities, to include interpretations at the personal, community, institutional and international levels. A strong focus of disaster and development studies surrounds the concept of 'avoidance': that is, prevention of disaster through development and better development through disaster reduction. Out of this stem a collection of key themes and conditions that will remain in high demand, including sustainability, coping, adaptation, resilience, security and wellbeing. Significant parts of the exercise of linking disaster and development have exposed the need for a way forward based on the identification of practical solutions backed up by theoretical and policy oriented insight. The field of disaster and development includes overlapping concerns, such as between science and technology, health security, advocacy and rights, risk governance, adaptive capacity and resilience. Substantial investments will be required in the coming years to ensure that work on these broadly integrated components is part of the solution, rather than just an endorsement of having recognised the nature of the problems.

Disaster reduction is fundamentally a development issue, whether in terms of dealing with risks associated with underdevelopment or any other form of inappropriate development. Development is also about learning and adapting to the risk of disaster. One of our key conclusions from the basic evidence provided is that disaster reduction requires a change in hearts and minds, together with adequate investment, if significant contributions are to be made to achieving MDGs and reducing the incidence of major catastrophes over the coming years (Figure 8.1). These include the adaptation of science and technology that can assist in improving the safety of industrialisation, reducing pollution and reducing our environmental footprint. Science and technology can also help reduce the impact of disaster by strengthening our defences to environmental hazards, be it through more resistant buildings, flood control, disease

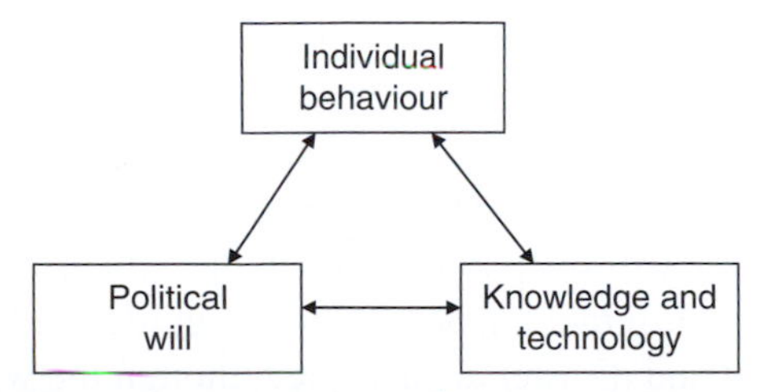

Figure 8.1 Underlying influences on successful disaster reduction initiatives

resistant crops and fire protection. Science and technology provide the development of more effective early warning systems, information technologies and the mathematics behind better predictive models. Better prediction of catastrophes improves predictability and reduces uncertainty. Reduced uncertainty as to how, why, where and when the next development issue will become a disaster, and the circumstances within which disasters may destroy development, improves the chances of finding solutions. The knowledge base of disaster management can be improved so that response systems are more efficient. Rapid communication of disaster events, identification of immediate needs, search and rescue and delivery of aid are all enhanced by our further investment in technological improvements in getting emergency assistance in the manner in which it is most needed.

Following disaster events it is also the further development of science and technology that provide the tools to understand what it was that happened and what went wrong. Disaster forensics is an increasingly important field, both in terms of accounting for the disasters of centuries gone by and in the area of recently occurred industrial or road accidents. The 'lessons learnt' exercises that follow most disasters draw on our scientific and investigative technologies. We have also seen that beyond hard scientific and technological approaches in disaster management there is a soft science approach that depends on our improved understanding of how to understand people, politics and cultures. The formulation of participatory or people centred approaches in disaster management and development work has opened up the possibility for improved communication, representation and rights in addressing disaster and development challenges around the world. It is clear that disasters impact disproportionately on some groups rather than others. We have learnt that disasters and development are interpretational spaces requiring multiple knowledges, perception studies and other cognitive behavioural work. The recent paper by Furedi (2007) illustrates that the debate on the cultural interpretation of 'disaster' is alive and well, particularly when arguing the merits of a focus on vulnerability or on resilience. The development of knowledge in this field also demands that we understand the 'non-scientific', which may be better described under the categories of culture, behaviour and politics. Moreover, others may be more comfortable exploring this in the light of an interdependent confluence of nature, culture and spirit. The meanings and realities of disaster for the majority of the world's population remain influenced by a faith of one sort or other. Different kinds of scientific evidence for the link between development and disaster are accepted, though the will to act in preventing disasters may be a function of multiple societal, economic and psychological influences.

Progress in risk reduction is dependent on development, with practice based disaster management requiring proactive and creative preventative engagement, rather than just a reaction to emergencies. It is the multiplication of positive or negative human behaviours that lie at the core of how we will avoid or reduce the risks of disasters. Put simply, those who are not part of the solution, through actions to conserve resources, avoid risky behaviour and resolve conflicts, are part of the contribution to disaster risk. In democratic societies we can to some extent influence outcomes by selecting leaders who care about disaster risk reduction and sustainable development. We choose how we will live and to what extent we hand down a less risky world to our descendents. Education and experience allow judgements about what is and what is not an unacceptable risk. Behaviour is also fundamentally affected by quality of life, including nutrition, leisure time, rights and inclusion in society. Where there is poverty and exclusion, people are less able to concentrate on the matters of risk reduction and planning for sustainable futures. It is not an effective approach to feed the poor with ideas alone, so progressing the disaster reduction agenda means tackling poverty head on.

To reduce poverty is to reduce disasters and to provide the opportunity for behavioural change. In this way, the link to politics and power is never far away. Hence the third part of our triangle of hope is to address political will. Where there is the political will, the context within which science, technology and behaviour change can advance is enhanced. The political will means that disasters and development are acknowledged as priority issues, and that investment is made at the right time to avert future catastrophes. The political context is also about creating an environment where the voice of those most at risk is heard, and facilitating the implementation of basic solutions to avert the worst forms of underdevelopment or exploitation. The political will of all nation-states is crucial in seeking to avoid conflicts, as allowing dangerous disagreements to arise in such a tightly packed world compromises the security of all. Political will and behavioural change in disaster reduction and development start at home, through raised awareness. They also involve the ability to admit to past mistakes and to define future development in ways that are sustainable.

The components of change for advancing disaster reduction are not separate compartments of activity, but are interrelated and dependent on each other. Thus, awareness, activism and taking responsibility for issues of global and local crisis are a behavioural matter that can influence the political will and provide the demand for better science and technology. Improved science and technology contribute to improved knowledge, evidence, understanding and lifestyle alternatives to affect behaviour, but it is important to recognise that indigenous and very localised practices can also be a strength. Both provide the means, through availability of new technology options, for the political will to change. As science and technology demonstrate they can come up with real solutions for industry, politicians can commit to early warning, to improved disaster response and to addressing sustainability issues. The political will can, conversely, drive science and technology in this field forward if government invests adequately in research and new and appropriate development initiatives. With the political will, communities and individuals can be persuaded to change their behaviour. Beyond encouraging changes in behaviour by choice, political will can drive legislation that effectively controls the threat of disasters. For example, if obesity is ranked one of the greatest disasters to threaten the UK in recent years, politicians can decide to ban fat inducing foodstuffs, fast food advertising and 'junk food' in school meals.

Sustainability and risk in the twenty-first century

There is a possibility of climate change affecting *en masse* the health and wellbeing of millions of poorer people as part of the world's 'development' process. However, if the industry of climate change serves to distract attention from the immediate issues of underdevelopment it will have contributed little. It may be that little was achieved at the World Conference on Environment and Development in Rio in 1992, but some of the issues in the way the environment is prioritised between North and South were exposed. Ten years on, at Johannesburg in 2002, the World Conference on Sustainable Development (WCSD) did at least put human interest first rather than the environment *per se*, and at Kobe in 2005 the World Conference on Disaster Reduction consolidated the start of a new institutional view on disaster risk reduction. The latter is based on the knowledge that it costs many times more to respond to disasters than to avoid them. This supports the ethos of climate change watchers that it is better to apply cautious risk management strategies with regard to emissions than to suffer the costs later. The precautionary principle more associated with climate debates was barely specifically mentioned at that event, but the entire disaster risk reduction approach now encapsulates it. These events take place alongside the coming into force of the Kyoto Protocol of the UN Framework Convention on Climate Change, and if the climate change debate has assisted in developing an emphasis on just some of the disaster risk reduction approach, it has helped rather than hindered this shift.

There has also been some progress in creating awareness that the creativity needed for disaster prevention often lies with local people in local settings dealing with local issues. This was a mainstream issue of the parallel discussions of non-governmental organisations, academics and other interested individuals engaging with Kobe. It has become blatantly obvious from several extreme disasters that communities' own knowledge, preparedness and actions are what saves lives. Some of that awareness was also apparent in the main UN forum through both North and South country statements. Acceptance of the strength of strategies embedded in the local, sensitive to people, place and time, is music to the ears of people who have worked for many years on the front line of community development programmes. However, it is also pertinent to point out that health and disaster risks and responses at the local level cannot be isolated from global insecurities through conflict, climate change or economic collapse, which are underlying international threats. Action points are therefore to create progress at local level amongst some of the more at-risk communities, whilst keeping global climate change firmly on the wider campaigning agenda. There is in this context a need to connect local risk reduction actions with wider advocacy work and macro-politics. The issues seem to be mainly about rights and representations in development that mitigate or reduce disaster risk, particularly for the poor and marginalised.

Some technologies, such as participatory action and research, are accepted as the way we can find out what people are doing and thinking with respect to health and other issues at the very local level (Cornwall *et al.* 2000). If these approaches are applied, people can use them at the local level as a means of grasping opportunities for representation and a better life. An approach that tends to be applied in the poorer parts of the world, the wider policy domain dominated by the rich and the powerful countries could learn from this in the interests of health and security. At a minimum, the public and private sectors need to respond by entering into commitments not to risk disrupting further the delicate balances between the wellbeing and ill health of the world's poor. Those refusing to be part of the agenda voluntarily arguably need to be put under pressure to respect the wider public's concerns for disaster reduction. However, rather than simply chastising business, industry and governments over a lack of action, it might be possible to gain further ground by providing educated evidence of the benefits that are to be had, economically, socially and environmentally, through investing in risk reduction. This agenda can progress even where there is only a small amount of evidence, though ongoing research is needed to more fully prove the point. The process of awareness building will, however, also now have to take place against a predictable onslaught of distractions from those who are keen to score points from gaps in knowledge in a game of uncertainty and risk.

Progress in disaster preparedness and control

Some of the progress is in improved information systems. Data on disasters, changes in development trajectory and reduction strategies have never been more available. Progress would appear to be in the recognition that prevention is better than cure, that disaster is not inevitable, and that through recognising and freeing up locally grounded routes to resilience people can become better protected. This has demanded a change in culture in disaster management, with the disaster and development logic providing a steer. Resilience has multiple dimensions, being applicable at personal, household, institutional, national and all levels. Paton and Johnston (2006) provide a useful collection of sketches on resilience, including resilience as capacity, the characteristics of a disaster resilient society, urban, social, ecological and economic resilience, and adaptation. There are also currently moves to apply resilience to the response and recovery end of disaster management. This is already prevalent with the use of resilience forums in the UK that are attended by emergency responders first and foremost. However, whilst resilience and also adaptation are key concepts, often with practical implications, this may be still just the early stages of considering people and disaster. Wisner *et al.* (2004: 372) advocated the development of a safety culture as the shift from a 'rudimentary level' of protection suggested by politicians and the media to an advanced level with public acceptance, education, laws, reduced risks and money allocated. A wide interagency project has gone as far as defining in some detail the characteristics of a disaster resilient community (Twigg 2007b). Communities are recognised as complex in this model, but community or systems resilience is understood to be:

- capacity to absorb stress or destructive forces through resistance or adaptation;
- capacity to manage, or maintain certain basic functions and structures, during disastrous events;
- capacity to recover or 'bounce back' after an event.

(Twigg 2007b: 6)

An expanded examination of disaster resilience was earlier provided by Manyena (2006), who endorses the sentiment that humans want to be more than just resilient. Whilst the recent focus on resilience has been very welcome in the disasters field, and a rich research topic in itself, inevitably it only provides some of the theoretical, methodological or policy solutions for disaster and development. Some of the shortcomings may be that it focuses more on what one needs to do as an individual, community or organisation to withstand disaster in the face of inevitable change. It also compliments the policy of adaptation, which itself is intensely political, as it begs the question of who is expected to adapt or to become resilient to whose threat. Theoretically, the extent to which a community might be encouraged to adapt or become 'resilient' might be an acceptance of the failure of government or the international community to reduce hazards. This has resonance in the context of climate change and complex political emergencies in particular. However, the point is introduced here purely for further thought as there are many nuances of interpretation to this argument. A further point might be that resilience is somewhat limited in development terms if interpreted along the lines of coping, whereby the objective is just to survive rather than to thrive. However, what is nonetheless very important about this approach is the recognition of the importance of the right type of enabling environment, within which communities can become resilient and achieve wellbeing.

People centred hazard and vulnerability reduction

One key way forward in addressing the subjectivity of what is and what is not vulnerability or resilience to disaster is to localise the assessment and management of hazards and risks within a community's own interpretations of them. This requires that less attention be given to what is or is not the ideal structure, and a greater focus on the developmental process of people taking ownership of ideas about disaster and development for themselves. An approach being used by several of the institutions engaging in projects with the Disaster and Development Centre is the establishment of community Risk and Resilience Committees (RRC), or similar. These have already been set up to manage health risks in Mozambique and Nepal to try to identify some of the optimum ways in which communities can self-organise to assess and manage risk and build resilience to multiple disaster types. The ones in Mozambique were specifically oriented towards reducing the impact of infectious disease.

The purpose of the Nepal version of RRC was:

- to protect communities from vulnerability and exposure to hazards;
- to inform the teaching and learning base with local knowledge;
- to provide a source of relevant information for circulation from community to community and to the wider governance context.

Activities have been as follows:

- to assess the manner in which communities build resilience;
- to assess how risk and resilience committees can improve wellbeing (i.e. not just to cope with disaster, but through sustainable development and wellbeing initiatives);
- to monitor changes in risks.

Some of the core ingredients of this process have been:

- identification of different interpretations and options for responding to risk and building on resilience;
- communication of risks and resilience;
- risk reduction;
- risk monitoring;
- evaluation and strengthening of risk governance.

Representation, rights and good governance in disasters

One of the key issues will remain how the vast majority of people who have yet to hear and interpret the language of disaster and development will be represented in a potentially more unstable world. Poverty, it has been said, can be not being able to read and write and to have others write about you. This is a sobering consideration as we near the end of this book. However, that people everywhere have rights not to be driven into increasingly more vulnerable situations requires a more simple acceptance of the norms of human decency. The equation will be hotly debated over the coming years and decades. Here, an initial sketch of some key points experienced in our fieldwork indicates that 'good disaster risk governance' should include the following:

- being informed by people's ongoing real or perceived threats;
- a practitioner orientation – guided by perpetual interpretation and review processes;

- proactive engagement with hazards, vulnerability, and coping to facilitate resilience through wellbeing;
- lessons learnt through evaluation before, during and after risk reduction activities;
- relative and localised knowledge made relevant through grounded research;
- people centred, driven and motivated disaster assessment that is multidisciplinary, integrated and perpetual;
- investment of finance and other resources into research, capacity building at all levels, and people's participation in decision making, all in the interests of implementing disaster avoidance.

Some of the common challenges that occur in this approach to risk and resilience for disaster reduction have been as follows:

- The language and meaning of the core concepts vary from culture to culture.
- Despite well-meaning research, it is not understood in depth what really makes a community more resilient in terms of the real and perceptual domains.
- Most success stories are part of something deeply embedded within a community that already exists before any external or formalised strategy for risk and resilience came on the agenda. The challenge in the face of new disaster threats can therefore be how to encourage multiplication and wider adoption of actions that already function rather than introduce new ones.
- Local risk management and resilience building may need to go further than adaptation of regular civil societal activities as there are contexts where no conventional community cohesion remains. This more rarely raises the possibility of the need to form new community structures to address risk and resilience activities.

Prospects in disaster risk reduction

There is an increase of written output based on the institutionalisation of disaster and development that has more recently culminated in a disaster risk reduction discourse. Meanwhile, its role amongst the daily realities of those living with disaster and development remains an avenue of exploration for the analyst, and of developmental entrepreneurship. The gap between researched and researcher may be worryingly large in terms of being able to offer solutions fast enough to improve the lives of those most at risk. Awareness about how best to describe age-old human exposure to hazards, or the progression of vulnerability, may have become an eloquent exercise, but threats to the lives of millions remain unremitting.

There is a need in the coming years to dig deep to address the demands created by our understanding of development induced disaster, and of disaster reduction for sustainable development. What is it that really drives people to the brink of desperation and beyond, but, more importantly, what is it that may sustain them and provide the human community with hope for the future? Many of those using this text are likely to be initiated into or in some way associated with the work of disaster and development studies. This entrance into the subject area is enhanced by some key ideas that people operating in contrasting contexts have adopted in a variety of ways. However, it is suggested that the initiated, as well as those already working with this material, maintain an open mind as to how best to conceptualise and implement disaster reduction for the wellbeing of the next period of life on earth. Where lessons have been 'learnt' the onus is to revisit them regularly, where inaction drives away from potential solutions we need to be activists, and where there is injustice we need a resolution to fairness.

From coping with catastrophe to living in security

As a somewhat idealised close to the book, one further model transition is summarised. It draws on much of the material in this text but does not claim to be comprehensive. However, in some ways it demonstrates a vision for the future and the transition from vulnerability to wellbeing, and it balances some of the key factors that may entail (Figure 8.2).

Some further notes on the human security aspect of this model are:

- Security lies at the intersection of environmental and human systems.
- In the contemporary context, we can consider security as the intersection of human and technological systems necessary for survival and wellbeing.
- Human security is core to (sustainable) development and disaster risk reduction, which can be assisted by technology.
- Security should first and foremost be acknowledged as a basic human need.
- Security is therefore a human right; people have rights, whilst machines, technologies, companies and corporations do not. However,

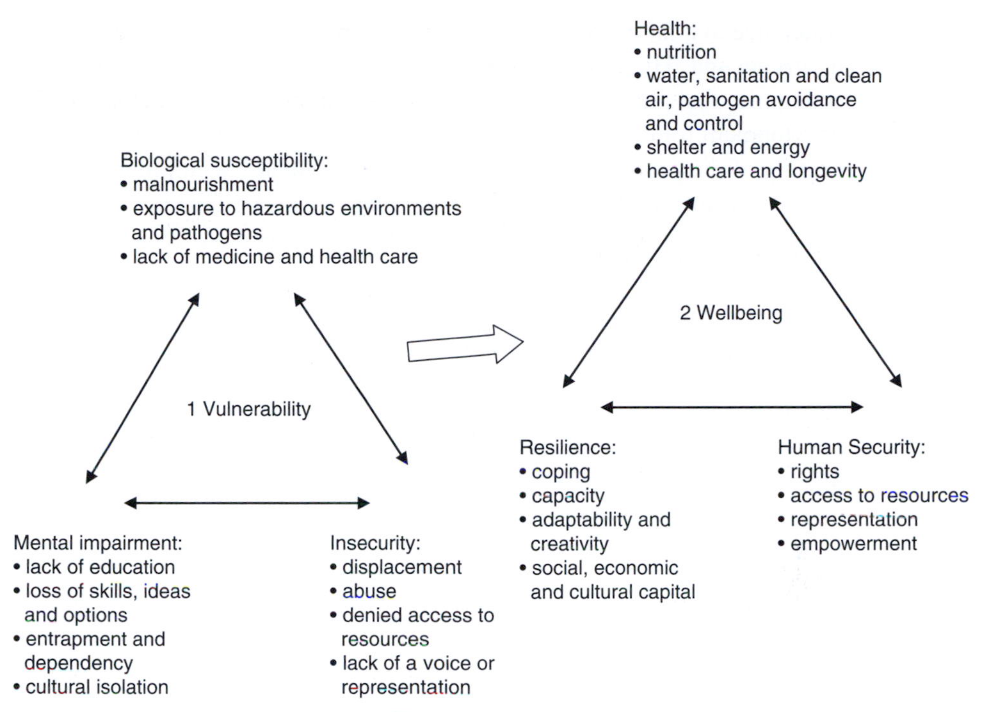

Figure 8.2 From integrated vulnerability to integrated wellbeing

technology and development are tools that can serve people in the upholding of their rights and security.

- The first principles of progressive disaster management perspectives are that early warning, risk management and preparedness bring security and are many times more cost-effective in dealing with disasters than responding to emergencies once disasters have struck.
- A second principle is that resilience to disasters must be embedded in multiple stakeholder or corporate interests in security. This is hopefully why we now have a Civil Contingencies Act and resilience forums in the case of the UK, and why community based disaster management is increasingly at the fore of the interests of emergency responders internationally. Both of these rationales appear to have been adopted by the Hyogo Accord for the current world strategy for disaster reduction (WCDR 2005).

An array of issues require further research, a few of which are pointed to here. They include the extent to which concepts of risk, resilience and wellbeing can be universalised in global strategies, given complex cultural interpretations of the same? With the majority of the world's adherence to a faith, we need to explore the role of the world's religions in this agenda. There is also a question mark over the extent to which economies that are already in decline can be oriented towards sustainable development for disaster reduction, when this is not understood as a priority of the present. There is a need to distinguish the optimum roles and responsibilities of individuals from those of institutions in implementing disaster risk reduction and sustainable development. What lies beyond the resilience agenda, and what is the right mix of education and sciences that is required by current and future generations to be able to address the increasing challenges of disaster and development? Questions remain as to how people mentally engage with disaster and development issues for the wellbeing of themselves and others. There are opportunities for the invention of new technologies to help strengthen infrastructures, improve early warning and emergency relief. The extent to which present net increases in wellbeing can offset future disaster threats and improve the effectiveness of disaster resilience and response is a further field of analysis. There is ultimately a need to know how one person's development opportunity can be safe from becoming someone else's disaster threat?

Discussion questions

1. What do you consider to be the meaning of disaster in terms of human development and security?
2. To what extent do you think disaster reduction can be improved by people changing their behaviour or by a reformulation of institutional structures?
3. Discuss the ways in which resilience might be best enhanced at the community level.
4. How might good risk governance become a universal right?
5. What gives you a sense of wellbeing, and how might you engage this in the interests of disaster reduction and sustainable development?

Further reading

Furedi, F. (2007) 'The Changing Meaning of Disaster', *Area* 39:4, pp. 482–489.

Paton, D. and Johnston, D. (eds) (2006) *Disaster Resilience: an integrated approach*, Springfield, IL: Charles Thomas Publishers.

Bibliography

Abrahamsson, H. and Nilsson, A. (1995) *Mozambique: the troubled transition - from socialist construction to free market capitalism*. London: Zed Books.

Adams, J. (1995) *Risk*, London: UCL Press.

Adams, W.M. (2001) *Green Development*, third edition, London: Routledge.

Adger, W.N., Paavola, J., Huq, S. and Mace, M.J. (eds) (2006) *Fairness in Adaptation to Climate Change*, London: MIT Press.

Alexander, D. (2000) *Confronting Catastrophe*, Harpenden: Terra Publishing.

Alexander, D. (2002) *Principles of Emergency Planning and Management*, Harpenden: Terra Publishing.

Alexander, D. (2008) 'Integrated Emergency Response: a resilience perspective', presentation to Dealing with Disasters Conference, Putting Resilience into Response Conference, 10 July, 2008.

Allen, T. and Thomas, A. (eds) (2000) *Poverty and Development into the Twenty-first Century*, Oxford: Oxford University Press.

ALNAP (2002) *Humanitarian Action: improving performance through improved learning*, Active Learning Network for Accountability and Performance in Humanitarian Action (ALNAP) Annual Review.

Amin, S. (1976) *Unequal Development*, Sussex: Harvester Press.

Anderson, M.B. (1999) *Do No Harm: how aid can support peace - or war?*, London: Lynne Rienner.

Anderson, M.B. and Woodrow, P.J. (1999) *Rising from the Ashes: development strategies in times of disaster*, London: Intermediate Technology Publications.

Annan, K. (2005) 'Towards a Culture of Peace. Letters to future generations', http://www.unesco.org/opi2/lettres/TextAnglais/AnnanE.html (accessed 24 January 2006).

Argenti-Pillen, A. (2003) *Masking Terror: how women contain violence in southern Sri Lanka*, Pennsylvania: University of Pennsylvania Press.

Arnstein, S.R. (1969) 'A Ladder of Citizen Participation', *Journal of the American Institute of Planners*, 35:4, pp. 216–224; reprinted in Gates, R.T. and Stout, F. (eds) (1969) *The City Reader*, second edition, London: Routledge.

Atampugre, N. (1997) 'Aid, NGO's and Grassroots Development: Northern Burkina Faso', *Review of African Political Economy* 71, pp. 57–73.

Ayres, R. (1995) *Development Studies: an introduction through selected readings,* Greenwich Readers, Greenwich: Greenwich University Press.

Balint, P.J. and Mashinya, J. (2008) 'CAMPFIRE through the Lens of the "Commons" Literature: Nyaminyami Rural District in post-2000 Zimbabwe', *Journal of Southern African Studies* 34:1, pp. 127–143.

Bankhoff, G., Frerks, G. and Hilhorst, D. (eds) (2004) *Mapping Vulnerability: disasters, development and people*, London: Earthscan.

Barnes, M.D., Hanson, C.L., Novilla, L.M.B., Meacham, A.I., McIntyre, E. and Erickson, B.C. (2008) 'Analysis of Media Agenda Setting during and after Hurricane Katrina: implications for emergency preparedness, disaster response, and disaster policy', *American Journal of Public Health* 98:4, pp. 604–610.

Barnett, T. and Whiteside, A. (2002) *AIDS in the Twenty-First Century: disease and globalization*, Basingstoke: Palgrave.

Barr, S. (2003) 'Strategies for Sustainability: citizens and responsible environmental behaviour', *Area* 35:3, pp. 227–240.

Barrett, H. (2009) *Health and Development*, Routledge Perspectives on Development Series, London: Routledge.

Barrow, C.J. (1997) *Environmental and Social Impact Assessment: an introduction*, London: Arnold.

Barrow, C.J. (1999) *Environmental Management: principles and practice*, London: Routledge.

Barrow, C.J. (2005) *Environmental Management and Development*, Routledge Perspectives on Development Series, London: Routledge.

Bates, I., Fenton, C., Gruber, J., Lalloo, D., Medina Lara, A., Squire, S.B., Theobald, S., Thomson, R. and Tolhurst, R. (2004a) 'Vulnerability to Malaria, Tuberculosis and HIV/AIDS Infection and Disease. Part 1: Determinants operating at individual and household level', *The Lancet of Infectious Diseases* 4:5, pp. 267–277.

Bates, I., Fenton, C., Gruber, J., Lalloo, D., Medina Lara, A., Squire, S.B., Theobald, S., Thomson, R. and Tolhurst, R. (2004b) 'Vulnerability to Malaria, Tuberculosis and HIV/AIDS Infection and Disease. Part 2: Determinants operating at environmental and institutional level', *The Lancet of Infectious Diseases* 4:6, pp. 368–375.

Beck, A., Emergy, G. and Greenberg, R.L. (1985) Anxiety, *Disorders and Phobias*, New York: Basic Books.

Beck, U. (1992) *Risk Society*: *towards a new modernity*, London: Sage.

Bell, S. and Morse, S. (1999) *Sustainability indicators: measuring the immeasurable,* Earthscan, London.

Ben-Haim, Y. (2006) *Info-gap Decision Theory: decisions under severe uncertainty*, second edition. London: Academic Press.

Benson, C. and Clay, E.J. (2004) *Understanding the Economic and Financial Impacts of Natural Disasters*, Disaster Risk Management Series No. 4, Washington, DC: World Bank.

Bhardwaj, S.M. and Rao, M.N. (1988) 'Regional development and seasonality of communicable diseases in rural Andhra Pradesh, India', *Social Science and Medicine* 26:1, pp. 15–24.

Bhatt, E.R. (1998) 'Women Victims' View of Urban and Rural Vulnerability', in Twigg, J. and Bhatt, M.R. (eds) *Understanding Vulnerability: South Asian perspectives,* Sri Lanka: ITDG, pp. 12–26.

Biel, R. (2000) *The New Imperialism: crisis and contradictions in North/South relations*, London: Zed Books.

Birkmann, J. (ed.) (2006) *Measuring Vulnerability to Natural Hazards: towards disaster resilient societies*, Tokyo: United Nations University Press.

Black, R. (1994) 'Environmental Change in Refugee-affected Areas of the Third World: the role of policy and research', *Disasters* 18:2, pp. 107–116.

Black, R. (1998) *Refugees, Environment and Development*, London: Longman.

Blackburn, J. and Holland, J. (eds) (1998) *Who Changes? Institutionalizing participation in development*, London: Intermediate Technology Publications.

Blaikie, P. (1999) 'A review of political ecology: issues, epistemology and analytical narratives'. *Zeitschriftfur Wirtschaftsgeographie* 43: pp. 131–147.

Blaikie, P.M. and Brookfield, H.C. (1987) *Land Degradation and Society*, London and New York: Methuen.

Blaikie, P., Cannon T., Davis I. and Wisner, B. (1994) *At Risk*, First edition, London: Routledge.

Bohle, H.G., Downing, T.E. and Watts, M.J. (1994) 'Climate Change and Social Vulnerability: toward a sociology and geography of food insecurity', *Global Environmental Change* 4:1, pp. 37–48.

Borroto, R.J. (1998) 'Global Warming, Rising Sea Level, and Growing Risk of Cholera Incidence: a review of the literature and evidence', *GeoJournal* 44:2, pp. v111–121.

Boserup, E. (1965) *The Conditions of Agricultural Growth: the economics of agrarian change under population pressure*. London: Allen and Unwin.

Bouma, M.J. and Pascual, M. (2001) 'Seasonal and Interannual Cycles of Endemic Cholera in Bengal 1891–1940 in Relation to Climate and Geography', *Hydrobiologia* 460, pp. 147–156.

Bracken, P.J., Giller, J.E. and Summerfield, D. (1995) 'Psychological Responses to War and Atrocity: the limitations of current concepts', *Social Science and Medicine* 40:8, pp. 1073–1082.

Brewer, A. (1990) *Marxist Theories of Imperialism: a critical survey*, London: Routledge.

Briceño, S. (2008) 'Today's Education for Tomorrow's Disaster Risk Reduction', in *Risk Wise*, www.tudor-rose.co.uk: Tudor Rose, pp. 14–17.

Brookfield, H. (1975) *Interdependent Development*, London: Methuen.

Brouwer, R. and Nhassengo, J. (2006) 'About Bridges and Bonds: community responses to the 2000 floods in Mabalane District, Mozambique', *Disasters* 30:2, pp. 234–255.

Brown, L.R. (2005) *Outgrowing the Earth: the food security challenge in an age of falling water tables and rising temperatures*, New York: W.W. Norton & Co.

Brown, V., Grootjans, J., Ritchie, J., Townsend, M. and Verrinder, G. (2005) *Sustainability and Health: supporting global ecological integrity in public health*, London: Earthscan.

Bryant, R.L. and Bailey, S. (1997) *Third World Political Ecology*, London: Routledge.

Buchanan, D.R. (2000) *An Ethic for Health Promotion: rethinking the sources of human well-being*, Oxford: Oxford University Press.

Buchanan-Smith, M. and Davies, S. (1995) *Famine Early Warning and Response: the missing link?*, London: Intermediate Technology.

Bull-Kamanga, L., Diagne, K., Lavell, A., Leon, E., Lerise, F., MacGregor, H., Maskrey, A., Pelling, M., Reid, H., Satterthwaite, D., Songsore, J., Westgate, K. and Yitambe, A. (2003) 'From Everyday Hazards to Disasters: the accumulation of risk in urban areas', *Environment and Urbanization* 15:1, pp. 193–203.

Burdge, R. and Vanclay, F. (1996). Social Impact Assessment: a contribution to the state of the art series', *Impact Assessment* 14(1): 59–86.

Cairncross, S. and Satterthwaite, D (1990) *The Poor Die Young: housing and health in third world cities*, London: Earthscan.

Carson, R. (1962) *Silent Spring.* St Louis: Houghton Mifflin.

Cernea, M.M. (1997) "The Risks and Reconstruction Model for Resettling Displaced Populations', *World Development* 25:10, pp. 1569–1588.

Cernea, M.M. (2000) 'Impoverishment Risks and Reconstruction: a model for population displacement and resettlement', in Cernea, M.M. and McDowell, C. (eds) *Risks and Reconstruction: experiences of resettlers and refugees*, Washington, DC: World Bank.

Cernea, M.M. (2005) 'Concept and Method: applying the IRR model', in Ohta, I. and Gebre, D. (eds) *Displacement Risks in Africa: refugees, resettlers and their host population*, Kyoto: Kyoto University Press and Trans Pacific Press, pp. 195–258.

Chambers, R. (1983) *Rural Development: putting the last first*, London: Longman Scientific and Technical.
Chambers, R. (1992) *Rural Appraisal: rapid, relaxed and participatory*, IDS Discussion Paper 311, Brighton: IDS.
Chambers, R. (1997) *Whose Reality Counts? Putting the first last*, London: Intermediate Technology Publications.
Chambers, R. (2005) *Ideas for Development*, London: Earthscan.
Chambers, R. and Conway, G. (1992) *Sustainable Rural Livelihoods: Practical concepts for the 21st century*, IDS Discussion Paper 296, Brighton: IDS.
Chambers, R., Longhurst, R. and Arnold, P. (eds) (1981) *Seasonal Dimensions to Rural Poverty*, Exeter: Francis Pinter.
Chowdhury, M.R. (2003) 'The Impact of "Greater Dhaka Flood Protection Project" (GDFPP) on Local Living Environment–the attitude of the floodplain residents', *Natural Hazards* 29:3, pp. 309–324.
Christian Aid (2007) *Human Tide: the real migration crisis*, a Christian Aid Report, London: Christian Aid.
Christie, F. and Hanlon, J. (2001) *Mozambique and the Great Flood of 2000*, African Issues, Oxford: James Currey.
Collins, A.E. (1996), 'The geography of Cholera', in Drasar, B.S. and Forrest, B.E. (eds) *Cholera and the Ecology of Vibrio Cholerae*, London: Chapman and Hall, pp. 255–294.
Collins, A.E. (1998) *Environment, Health and Population Displacement: development and change in Mozambique's diarrhoeal disease ecology*, Making of Modern Africa Series, Aldershot: Ashgate.
Collins, A.E. (2001) 'Health Ecology, Land Degradation and Development', *Land Degradation and Development* 12:3, pp. 237–250.
Collins, A.E. (2002) 'Health Ecology and Environmental Management in Mozambique', *Health and Place* 8:4, pp. 263–272.
Collins, A.E. (2003) 'Vulnerability to Coastal Cholera Ecology', *Social Science and Medicine*, 57, pp. 1397–1407.
Collins, A.E. and Lucas, M. (2002) '*O Que Aconteceu?*' (What happened?), Report on environmental health preparedness, response and lessons learnt from recent flooding emergency in Mozambique, Maputo: Ministry of Health, Government of Mozambique.
Collins, A.E., Lucas, M.E., Islam, M.S. and Williams, L.E. (2006) 'Socio-economic and Environmental Origins of Cholera Epidemics in Mozambique: guidelines for tackling uncertainty in infectious disease prevention and control', *International Journal of Environmental Studies*, Special Issue on Africa, 63:5, pp. 537–549.
Collinson, S. (ed.) (2003) *Power, Livelihoods and Conflict: case studies in political economy analysis for humanitarian action*, HPG report 13, London: ODI.
Colwell, R. (2002) 'A Voyage of Discovery: cholera, climate and complexity', *Environmental Microbiology* 4:2, pp. 67–69.
Colwell, R.R., and Spira, W.M. (1992) 'The ecology of Vibrio Cholerae', in Barua, D. and Greenough, W.B. III (eds) *Cholera*, New York: Plenum Medical Book Company, pp. 107–127.
Commission on Human Security (2003) *Human Security Now*, New York: Commission on Human Security.
Commonwealth Secretariat (2002) *Gender Mainstreaming in HIV/AIDS: taking a multisectoral approach*, London: Commonwealth Secretariat.
Connolly, M.A., Gayer, M., Ryan, P.S., Spiegel, P. and Heymann, D.L. (2004) 'Communicable Diseases in Complex Emergencies: impact and challenges', *Lancet* 364, pp. 1974–1983.
Cook, A., Watson, J., van Buynder, P., Robertson, A. and Weinstein, P. (2008) '10th Anniversary Review: natural disasters and their long-term impacts on the health of communities', *Journal of Environmental Monitoring* 10, pp. 167–175.

Cooke, B. and Kothari, U. (eds) (2001) *Participation: the new tyranny?*, London: Zed Books.

Cornwall, A., Lucas, H. and Pasteur, K. (eds) (2000) 'Accountability through Participation: developing workable partnership models in the health sector,' *IDS Bulletin* 31:1, pp. 1–13.

Cox, R.S., Long, B.C., Jones, M.I. and Handler, R.J. (2008) 'Sequestering of Suffering - critical discourse analysis of natural disaster media coverage', *Journal of Health Psychology* 13:4, pp. 469–480.

Cracknell, B.E. (2000) *Evaluating Development Aid: issues, problems and solutions*, London: Sage.

CRED (Center for Research on the Epidemiology of Disaster) (2006) *Annual Disaster Statistical Review: the numbers and trends 2005*, Melin Belgium: Jacoffset Printers.

CRED (Centre for Research on the Epidemiology of Disaster) (2008) *Annual Disaster Statistical Review: the numbers and trends 2007*, Melin, Belgium: Jacoffset Printers.

Crisp, J. (2005) 'No Solutions in Sight: the problem of protracted refugee situations in Africa', in Ohta, I. and Gebre, D. (2005) *Displacement Risks in Africa: refugees, resettlers and their host population*, Kyoto: Kyoto University Press and Trans Pacific Press, pp. 17–52.

Croner (2002) *A-Z Essentials: environmental management,* London: Croner CCH Ltd.

Cuny, F. (1983) *Disasters and Development*, Oxford: Oxford University Press.

Curto de Casas, S.I. (1994) 'Health Care in Latin America', in Phillips, D.R. and Verhasselt, Y. (eds) *Health and Development*, London: Routledge, pp. 234–248.

Curtis, S. and Taket, A. (1996) *Health and Societies: changing perspectives*, London: Arnold.

Davis, A.P. (1996) 'Targeting the Vulnerable in Emergency Situations: who is vulnerable?', *Lancet* 348:9031, pp. 868–871.

Davis, I. (2004) 'Progress in Analysis of Social Vulnerability and Capacity', in Bankhoff, G., Frerks, G. and Hilhorst, D. (eds) *Mapping Vulnerability: disasters, development and people*, London: Earthscan, pp. 128–144.

DDC (Disaster and Development Centre) (2008) *Five Year Plan 2007–12*, Northumbria University.

De Waal, A. (1997) *Famine Crimes: politics and the disaster relief industry in Africa*, African Issues, London: James Currey.

Desjarlais, R., Eisenberg, L., Good, B. and Kleinman, A. (1995) *World Mental Health: problems and priorities in low-income countries*, New York: Oxford University Press.

Devereux, S. (1993) *Theories of Famine*, London: Harvester Wheatsheaf.

Devereux, S. and Maxwell, S. (eds) (2001) *Food Security in Sub-Saharan Africa*, London: ITDG Publishing.

DFID (Department for International Development) (1997) *Eliminating World Poverty: a challenge for the 21st century*, White Paper on International Development, London: HMSO.

DFID (1999) *Sustainable Livelihoods Guidance Sheets*, London: DFID.

DFID (2000) *Sustainable Livelihoods Guidance Sheets*, London: DFID.

DFID (Department for International Development) (2005) *Disaster Risk Reduction: a development concern*, London: DFID.

DFID (Department for International Development) (2006) *Reducing the Risk of Disasters: helping to achieve sustainable poverty reduction in a vulnerable world: a DFID policy paper*, London: DFID.

Diacon, D. (1997) 'Typhoon Resistant Housing for the Poorest of the Poor in the Philippines'. in Awotona, A. *Reconstruction after Disaster: issues and practices*, Aldershot: Ashgate, pp. 130–147.

Dixon, J.A., Scura, L., Carpenter, R. and Sherman, P. (1994) *Economic Analysis of Environmental Impacts*, London: Earthscan in association with the Asian Development Bank and the World Bank.

Dodd, R. and Cassels, A. (2006) 'Health, Development and the Millennium Development Goals', *Annals of Tropical Medicine and Parasitology*, 100: 5/6, pp. 379–387.

Dolce, A. and Ricciardi, M. (2007) 'Impact of Psychological Risk Factors on Disaster Rescue Operations: the case of Italian volunteers', *Disasters* 31:1, pp. 91–103.

Dombrowsky, W.R. (1998) 'Is a Disaster What We Call a "Disaster"?', in Quarantelli, E.L. (ed) (1998) *What Is a Disaster? Perspectives on the question*, London: Routledge, pp. 19–30.

Donini, A., Minear, L., Smillie, I., van Baarda, T. and Welch, A.C. (2005) *Mapping the Security Environment: understanding the perceptions of local communities, peace support operations and assistance agencies*, Medford MA: Feinstein International Famine Center. Tufts University.
Douglas, M. (1996) *Risk Acceptability According to the Social Sciences*, London: Routledge.
Doyal, L. (1987) *The Political Economy of Health*, London: Pluto Press.
Drasar, B.S. and Forrest, B.D. (eds) (1996) *Cholera and the Ecology of Vibrio Cholerae*, London: Chapman & Hall.
Ecologist (1992) 'Whose Common Future?' *Ecologist* 22: 4 (special issue).
Ecologist (1993) *Whose Common Future?: Reclaiming the commons*, London: Routledge.
Edgeworth, R. and Collins, A.E. (2006) 'Self-care as a Response to Diarrhoea in Rural Bangladesh: empowered choice or enforced adoption?', *Social Science and Medicine* 63, pp. 2,686–2,697.
Edwards, F. (1996) *NGO Performance: what breeds success*?, London: Intermediate Technology Publications.
Ehrlich, P.R. (1962) *The Population Bomb*, New York: Ballantine Books.
Ehrlich, P.R. and Ehrlich, A.H. (1990) *The Population Explosion*, New York: Simon and Schuster.
Ellen, R.F. (1984) *Ethnographic Research: a guide to general conduct*, London: Academic Press.
Elliot, J.E. (2006) *An Introduction to Sustainable Development*, third edition, London: Routledge.
Eshuis, J. and Manschot, P. (1992) *Communicable Diseases: a manual for rural health workers,* Nairobi: African Medical and Research Foundation.
Estrella, M. (ed.) (2000) *Learning from Change: issues and experiences in participatory monitoring and evaluation*, London: ITP.
FAOSTAT (2008) Statistical Database, Food and Agriculture Organization of the United Nations, http://faostat.fao.org (accessed 30 July 2008).
Few, R. and Matthies (eds) (2006) *Flood Hazards and Health: responding to present and future risks*, London: Earthscan.
Few, R. (2006) 'Flood Hazards, Vulnerability and Risk Reduction', in *Flood Hazards and Health: responding to present and future risks*, London: Earthscan, pp. 8–27.
Few, R. (2007) 'Health and Climatic Hazards: framing social research on vulnerability, response and adaptation', *Global Environmental Change* 17:2, pp. 281–295.
Fordham, M. (2003) 'Gender, Disaster and Development: the need for integration', In Pelling, M. (ed) (2003) *Natural Disasters and Development in a Globalising World*, London: Routledge, pp. 57–74.
Forsyth, T.J. (2002) *Critical Political Ecology*, London: Routledge.
Fowler, A. (1997) *Striking a Balance: a guide to enhancing the effectiveness of non-governmental organizations in international development*, London: Earthscan.
Francis, P. (2001) 'Participatory Development at the World Bank: the primacy of process', in Cooke, B. and Kothari, U. (eds) *Participation: the new tyranny?*, London: Zed Books, pp. 72–87.
Frank, G.A. (1966) 'The Development of Underdevelopment', *Monthly Review* 18:4, pp. 17–31.
Frank, G.A. (1975) *On Capitalist Underdevelopment*, Oxford: Oxford University Press.
Freire, P. (1972) *Pedagogy of the Oppressed*, Harmondsworth: Penguin Books.
Frerks, G.E., Kliest, T.J., Kirkby, S.J., Emmel, N.D., O'Keefe, P. and Convery, I. (1995) 'A Disaster Continuum?', *Disasters* 19:4, pp. 326–366.
Furedi, F. (2007) 'The Changing Meaning of Disaster', *Area* 39:4, pp. 482–489.
Gebre, Y.D. (2005) 'Promises and Predicaments of Resettlement in Ethiopia', in Ohta, I. and Gebre, D. (eds) *Displacement Risks in Africa: refugees, resettlers and their host population*, Kyoto: Kyoto University Press and Trans Pacific Press, pp. 359–384.
Gebre, Y.D. and Ohta, I. (2005) 'Introduction: displacement in Africa: conceptual and practical concerns', in Ohta, I. and Gebre, Y.D. (eds) *Displacement Risks in Africa: refugees, resettlers and their host population*, Kyoto: Kyoto University Press and Trans Pacific Press, pp. 1–14.
George, S. (1992) *The Debt Boomerang*, London: Pluto Press.

Goma Epidemiology Group (1995) 'Public Health Impact of Rwandan Refugee Crisis: what happened in Goma, Zaire, in July, 1994?', *Lancet* 345, pp. 339–344.

Goodhand, J., Hulme, D. and Lewer, N. (2000) 'Social Capital and the Political Economy of Violence: a case study of Sri Lanka', *Disasters* 24:4, pp. 390–406.

Gosling, L. and Edwards, M. (2003) *Toolkits: a practical guide to planning, monitoring, evaluation and impact assessment*, London: Save the Children.

Gray, A. (2001) *World Health and Disease*, third edition, Health and Disease Series, Milton Keynes: Open University Press.

Green, R.H. (1992) 'The Four Horseman Ride Together: scorched fields of war in Southern Africa', paper presented at Refugees Studies Programme, Oxford, 11 November.

Guest, E. (2003) *Children of AIDS: Africa's orphan crisis*, London: Pluto Press.

Guha-Sapir, D and Below, R. (2006) *The Quality and Accuracy of Disaster Data: a comparative analyses of three global datasets*, Provention Consortium. World Bank Disaster Management Facility, Washington, DC: World Bank.

Haines, A. and Patz. J. (2004) 'Health Effects of Climate Change', *Journal of the American Medical Association* 291:1, pp. 99–103.

Hanlon, J. (1991) *Mozambique: who calls the shots?*, London: James Currey.

Hanlon, J. (1996) *Peace without Profit: how the IMF blocks rebuilding in Mozambique*, London: James Currey.

Hannigan, J.A. (1995) *Environmental Sociology: a social constructionist perspective*, London: Routledge.

Hardin, G. (1968) 'The Tragedy of the Commons', *Science* 162, pp. 1243–1248.

Hardin, G. (1974) *The Ethics of a Lifeboat*, Washington, DC: American Association for the Advancement of Science.

Hardoy, J.E., Mitlin, D. and Satterthwaite, D. (1992) *Environmental Problems in Third World Cities*, London: Earthscan.

Harrell-Bond, B., Voutir, E. and Leopold, M. (1992) 'Counting the Refugees: gifts, givers, patrons, clients', *Journal of Refugee Studies* 5:3–4, pp. 205–225.

Harremoës, P., Gee, D., MacGarwin, M., Stirling, A., Keys, J., Wynne, B. and Guedes, V. (eds) (2002) *The Precautionary Principle in the 20th Century: late lessons from early warnings*, London: Earthscan and European Environment Agency.

Hatzius, T. (1996) *Sustainability and Institutions: catchwords or new agenda for ecologically sound development*?, IDS Working Paper 48, Brighton: Institute of Development Studies.

Hayter, T. (1981) *The Creation of World Poverty: an alternative view to the Bandt Report*, London: Pluto.

Helmer, M. and Hilhorst, D. (2006) 'Natural Disasters and Climate change', *Disasters* 30:1, pp. 1–4.

Hewitt, K. (1997) *Regions of Risk: a geographical introduction to disasters*, Harlow: Longman.

Hewitt, K. (1998) 'The Social Construction of Disaster', in Quarantelli, E.L. (ed) (1998) *What Is a Disaster? Perspectives on the question*, London: Routledge, pp. 75–91.

Hodgkinson, P.E. and Stewart, M. (1998) *Coping with Catastrophe: a handbook of post-disaster psychosocial aftercare*, London: Routledge.

Home Office (1994) *Dealing with Disaster*, second edition, London: HMSO.

Hoogvelt, A. (2001) *Globalization and the Postcolonial World: the new political economy of development*, Basingstoke: Palgrave.

HPA (Health Protection Agency) (2003) presentation by Dr Nigel Calvert, HPA consultant in communicable disease control to Disaster Management Programme, Northumbria University, 17 March.

HPA–CDR (Health Protection Agency–Communicable Disease Report) (2005) 'Creutzfeldt-Jakob Disease (CJD) Update Report', *HPA - CDR Weekly* 15:6.

HPA (Health Protection Agency) (2005) *Health Protection Agency - influenza pandemic contingency plan*, London: HPA.

Hubbard, M. (1995) *Improving Food Security: a guide for rural development managers*, London: I.T. Publications.

Hulme, D. and Edwards, M. (eds) (1997) *NGOs, States and Donors: too close for comfort?*, Basingstoke: Macmillan.

Hyslop, M. (2007) *Critical Information Infrastructure: resilience and protection*, New York: Springer.

IFRC (International Federation of Red Cross and Red Crescent Societies) (2001) *World Disasters Report 2001*: Focus on Recovery. Geneva: IFRC.

IFRC (International Federation of Red Cross and Red Crescent Societies) (2002) *World Disasters Report 2002*: Focus on Reducing Risk, Geneva: IFRC.

IFRC (International Federation of Red Cross and Red Crescent Societies) (2004) *World Disasters Report 2004*: Focus on Community Resilience, Geneva: IFRC.

IFRC (International Federation of Red Cross and Red Crescent Societies) (2005) *World Disasters Report 2005*: Focus on Information in Disasters, Geneva: IFRC.

IFRC (International Federation of Red Cross and Red Crescent Societies) (2006) *World Disasters Report* 2006: Focus on Neglected Disasters, Geneva: IFRC.

IFRC (1996) *Reducing Risk: participatory learning activities for disaster mitigation in Southern Africa*, Geneva IFRC.

IASC (Inter-Agency Standing Committee) (2007) *IASC Guidelines on Mental Health and Psychosocial Support in Emergency Settings*, Geneva: IASC.

IDS (Institute of Development Studies) (1995) *Confronting Famine in Africa*, IDS Policy Briefing Issue 3, Brighton: Institute of Development Studies.

IPCC (Intergovernmental Panel on Climate Change) (1995) *Climate Change 1995: impacts, adaptations and mitigation of climate change: scientific-technical analyses*. Contribution of Working Group II to the Second Assessment Report of the Intergovernmental Panel on Climate Change, Cambridge: Cambridge Press.

IPCC (Intergovernmental Panel on Climate Change) (2007) *Climate Change 2007: the physical science basis: summary for policymakers*, Contribution of Working Group I to the Fourth Assessment Report of the Intergovernmental Panel on Climate Change, Geneva: WMO and UNEP .

Islam, M.S., Drasar, B.S. and Bradley Sack, R. (1993) 'The Aquatic Environment as a reservoir of *Vibrio cholerae*: a review', *Journal of Diarrhoeal Disease Research* 11:4, pp. 197–206.

Izadkhah, Y.O. and Mahmood, H. (2005) 'Towards Resilient Communities in Developing Countries through Education of Children for Disaster Preparedness', *International Journal of Emergency Management* 2:3, pp. 138–148.

Jones, L. (2004) *Then They Started Shooting: growing up in wartime Bosnia*, Cambridge, MA: Harvard University Press.

Jones, K. and Moon, G. (1996) *Health, Disease and Society: an introduction to medical geography*, London: Routledge.

Kaldor, M. (2007) *Human Security*, Cambridge: Polity Press.

Kalipeni, E. and Oppong, J. (1998) 'The Refugee Crisis in Africa and Implications for Health and Disease: a political ecology approach', *Social Science and Medicine* 46:12, pp. 1637–1653.

Kalipeni, E. (2000) 'Health and Disease in Southern Africa: a comparative and vulnerability perspective', *Social Science and Medicine* 50:7, pp. 965–983.

Kasperson, J.X. and Kasperson, R.E. (eds) (2001) *Global Environmental Risk*, London: Earthscan.

Kellow, A. (1999) *International Toxic Risk Management: ideals, interests and implementation*, Cambridge: Cambridge University Press.

Keynes, J.M. (1936) *The General Theory of Employment, Interest and Money*, London: Macmillan.

King, D. and Cottrell, A. (eds) (2007) *Communities Living with Hazards*, Queensland: Centre for Disaster Studies, James Cook University.

Kovats, R.S., and Bouma, M.J. and Haines, A. (2003) 'El Nino and Health', *The Lancet* 362:9394, pp. 1481–1489.

Kurimoto, E. (2005) 'Multidimensional Impact of Refugees and Settlers in the Gambela Region, Western Ethiopia', in Ohta, I. and Gebre, D. (eds) *Displacement Risks in Africa: refugees, resettlers and their host population*, Kyoto: Kyoto University Press and Trans Pacific Press, pp. 338–358.

Lafond, A. (1995) *Sustaining Primary Health Care*, London: Intermediate Technology Publications.

Lama, J.R. Seas, C.R., León – Barúa, R., Gotuzzo, E. and Sack, R.B. (2004) 'Environmental Temperature, Cholera and Acute Diarrhoea in Adults in Lima, Peru', *Journal of Health, Population and Nutrition* 22:4, pp. 399–403.

Lappe, F.M., Collins, J. and Rosset, P. (1998) *World Hunger: 12 myths*, London: Earthscan.

Larkin, M. (1998) 'Global Aspects of Health and Health Policy in Third World Countries'. in Kiely, R. and Marfleet, P. (eds) *Globalisation and the Third World*, London: Routledge.

Larson, K.L. and Lach, D. (2008) 'Participants and Non-participants of Place-based Groups: an assessment of attitudes and implications for public participation in water resource management', *Journal of Environmental Management* 88:4, pp. 817–830.

Laws, S. (2003) *Research for Development: a practical guide*, London: Sage.

Learmonth, A. (1988) *Disease Ecology*, Oxford: Blackwell.

Lee, N. and George, C. (2000) *Environmental Assessment in Developing and Transitional Countries*, London: Wiley.

Leon, D.A., Walt, G. and Gilson, L. (2001) 'International Perspectives on Health Inequalities and Policy', *British Medical Journal* 322, pp. 591–594.

Lewis, J. (1999) *Development in Disaster-prone Places: studies of vulnerability*, London: ITP.

Linnerooth-Beyer, J., Löfstedt, R.E and Sjöstedt, G. (eds) (2001) *Transboundary Risk Management*, London: Earthscan.

Linnerooth-Beyer, J., Mechler, R. and Pflug, G. (2005) 'Refocusing Disaster Aid', *Science* 309, pp. 1,044–1,046.

Lipp, E.K, Huq, A. and Colwell, R.R. (2002) 'Effects of Global Climate on Infectious Disease: the cholera model', *Clinical and Microbiological Review* 15:4, pp. 757–770.

Liu, C. (2003) 'The Battle against SARS: a Chinese story', *Australian Health Review* 26:3, pp. 3–13.

McEntire, D. Fuller, C., Johnston, C. and Weber, R. (2002) 'A Comparison of Disaster Paradigms: the search for a holistic policy guide', *Public Administration Review* 62:3, pp. 267–281.

McEntire, D.A. (2004) 'Development, Disasters and Vulnerability: a discussion of divergent theories and the need for their integration', Disaster Prevention and Management 13:3, pp. 193–198.

McEntire, D. (2007) *Disaster Response and Recovery*, Moboken, NJ: Wiley.

McMurray, C. and Smith, R. (2001) *Diseases of Globalization: socioeconomic transitions and health*, London: Earthscan.

Macrae, J. (1995) *Dilemmas of 'Post'-Conflict Transition: lessons from the health sector*, Relief and Rehabilitation Network Paper 12, London: ODI.

Manyena, B. (2006) 'The Concept of Resilience Revisited', *Disasters* 30:4, pp. 433–450.

Maxwell, S. (2001) 'The Evolution of Thinking about Food Security', in Devereux, S. and Maxwell, S. (eds) *Food Security in Sub-Saharan Africa*, London: ITDG, pp. 13–32.

May, J.M. (1958) *The Ecology of Human Disease*, New York: M.D. Publications.

May, J.M. (1960) *Disease Ecology*, New York: Hafner.

Mayer, J.D. (1996) 'The Political Ecology of Disease as One New Focus for Medical Geography', *Progress in Human Geography* 20:4, pp. 441–456.

Mbakem-Anu, E. (2007) 'Displacement, Livelihoods and Sustainable Development', PhD thesis, Northumbria University.

Meade, M.S. (1977) 'Medical geography as Human Ecology: the dimension of population movement', *Geographical Review* 67:4, pp. 377–399.

Meadows, D.H., Meadows, D.L., Randers, J. and Behrens, W.W. III, (1972) *The Limits to Growth* (a report for the Club of Rome's project on the predicament of mankind), New York: Universal Books.

Meadows, D.H., Meadows, D.L. and Randers, J. (1992) *Beyond the Limits: global collapse or sustainable future?*, London, Earthscan.

Middleton, N. and O'Keefe, P. (1998) *Disaster and Development: the politics of humanitarian aid*. London: Pluto.

Middleton, N. and O'Keefe, P. (2001) *Negotiating Poverty: new directions, renewed debate*, London: Pluto.

Middleton, N. and O'Keefe, P. (2003) *Rio Plus Ten: politics, poverty and the environment*, London: Pluto.

Millennium Ecosystem Assessment (2005) *Ecosystems and Human Well-being: synthesis*, Washington, DC: Island Press.

Mohan, G. (2002) 'Participatory Development in Practice', in Desai, V. and Potter, R.B. (eds) *The Companion to Development Studies*, London: Arnold.

Momsen, J.H. (2003) *Gender and Development*, London: Routledge.

Montano, D. (1986) 'Predicting and Understanding Influenza Vaccination Behaviour: alternatives to the health belief model', *Medical Care* 5, pp. 438–453.

Mowforth, M. (2000) *Storm warnings Hurricanes Georges and Mitch and the lessons for development*, CIIR Briefing Document, London: Catholic Institute for International Relations.

MSF (Médicins Sans Frontières) (1997) *Refugee Health: an approach to emergency situations*, London: Macmillan Education Ltd.

Neal, D.M. and Phillips, B.D. (1995) 'Effective Emergency Management - reconsidering the bureaucratic approach', *Disasters* 19:4, pp. 327–337.

Neefjes, K. (2000) *Environments and Livelihoods – strategies for sustainability*, London: Oxfam and Redwood Books.

Noji, E.K. (ed) (1997) *The Public Health Consequences of Disasters*, Oxford: Oxford University Press.

Nordberg, E. (ed) (1999) *Communicable Diseases*, third edition, African Medical and Research Foundation (AMREF), Rural Health Series No.7, Nairobi: AMREF.

Norwegian Refugee Council (2002) *Internally Displaced People: a global survey*, London: Earthscan.

O'Brien, G. and Read, P. (2005) 'Future UK Emergency Management: new wine, old skin?', *Disaster Prevention and Management* 14:3, pp. 353–361.

O'Brien, G., O'Keefe, P., Rose, J. and Wisner, B. (2006) 'Climate Change and Disaster Management', *Disasters* 30:1, pp. 64–80.

O'Riordan, T. (1976) *Environmentalism*, London: Pion.

Oakley, P. and Clayton, A. (2000) *The Monitoring and Evaluation of Empowerment: a resource document*, Oxford: INTRAC.

Oakley, P., Pratt, B. and Clayton, A. (1998) *Outcomes and Impact: evaluating change in social development*, Oxford: INTRAC.

OECD–DAC (2001) Poverty *Reduction: the DAC guidelines*, Paris: OECD–DAC.

Ohta, I. (2005) 'Multiple Socio-economic Relationships Improvised between the Turkana and Refugees in Kakuma Area, Northwestern Kenya', in Ohta, I. and Gebre, D. (eds) *Displacement Risks in Africa: refugees, resettlers and their host population*, Kyoto: Kyoto University Press and Trans Pacific Press, pp. 315–337.

Ohta, I. and Gebre, T.D. (eds) (2005) *Displacement Risks in Africa: refuges, resettles and their host population*, Kyoto University Press and Trains Pacific Press.

Oliver-Smith, A. (1996) "Anthropological Research on Hazards and Disasters' *Annual Review of Anthropology* 25, pp. 303–328.

Oliver-Smith, A. (2004) 'Theorizing vulnerability in a Globalized World: a political ecological perspective', in Bankoff, G., Frerks, G. and Hilhorst, D. (eds) *Mapping Vulnerability: disasters, development and people*, London: Earthscan, pp. 10–24.

Oxfam (1995) *The Oxfam Handbook of Development and Relief*, Oxford: Oxfam.

Özerdem, A. (2003) 'Disaster as manifestation of Unresolved Development Challenges: the Marmara earthquake, Turkey', in Pelling, M.(ed.) *Natural Disasters and Development in a Globalising World*, London: Routledge. pp. 199–213.

Pascual, M., Rodó, X., Ellner, S.P., Colwell, R. and Bouma, M.J. (2000) 'Cholera Dynamics and El Niňo Southern Oscillation', *Science* 289:5485, pp. 1766–1775.

Pascual, M., Bouma, M.J.and Dobson, A.P. (2002) 'Cholera and Climate: revisiting the quantitative evidence', *Microbes and Infection* 4:2, pp. 237–245.

Paton, D. and Auld, T. (2006) 'Resilience in Emergency Management: managing the flood'. in Paton, D. and Johnston, D. (eds) *Disaster Resilience: an integrated approach*, Springfield, IL: Charles Thomas Publishers, pp. 267–287.

Paton, D. and Johnston, D. (eds) (2006) *Disaster Resilience: an integrated approach*, Springfield, IL: Charles Thomas Publishers.

Patz, J.A. and Kovats, R.S. (2002) 'Hotspots in Climate Change and Human Health', *British Medical Journal* 325, pp. 1,094–1,098.

Pearlman, L.A. and Saakvitne, K.W. (1995) *Trauma and the Therapist*, New York: Norton.

Peet, R. and Watts, M. (eds) (2004) *Liberation Ecologies: environment, development and social movements*, second edition, London: Routledge.

Pelling, M. (2003) *The Vulnerability of Cities: natural disasters and social resilience*, London: Earthscan.

Pelling, M. (2003b) 'Emerging Concerns', in Pelling, M. (ed.) *Natural Disasters and Development in a Globalising World*, London: Routledge, pp. 233–243.

Pelling, M. (2007) 'Learning from Others: the scope and challenges for participatory disaster risk assessment', *Disasters* 31, pp. 373–385.

Pepper, D. (1984) *The Roots of Modern Environmentalism*, London: Croom Helm.

Pepper, D. (1996) *Modern Environmentalism: an introduction*, London: Routledge.

Philips, D.R. and Verhasselt, Y. (1994) 'Introduction: health and development', in Phillips, D.R. and Verhasselt, Y. (eds) *Health and Development*, London: Routledge.

Pirotte, C., Husson, B. and Grunewald, F. (1999) *Responding to Emergencies and Fostering Development: the dilemmas of humanitarian aid*, London: Zed Books.

Platt, R.H. (1999) *Disasters and Democracy: the politics of extreme natural events*, Washington DC: Island Press.

Popper, K. (2002) [1959] *The Logic of Scientific Discovery,* New York: Routledge Classics.

Porteous, D. (1989) *Planned to Death: the annihilation of a place called Howdendyke*, Manchester: Manchester University Press.

Prothero, R.M. (1977) 'Disease and Human Mobility: a neglected factor in epidemiology', *International Journal of Epidemiology* 6, pp. 259–267.

Prothero, R.M. (1994) 'Forced Movements of Population and Health Hazards in Tropical Africa', *International Journal of Epidemiology* 23:4, pp. 657–664.

Pupavac, V. (2001) 'Therapeutic Governance: psycho-social intervention and trauma risk management', *Disasters* 25:4, pp. 358–372.

Quarantelli, E.L. (ed.) (1998) *What is a Disaster? Perspectives on the question*, London: Routledge.

Reed, D. (ed.) (1993) *Structural Adjustment and the Environment*, London: Earthscan.

Ribot, J. (2002) 'Democratic Decentralisation of Natural Resources: institutionalising popular participation'. World Resources Institute, www.wri.org/governance/pubs_description.cfm?pid=3767 (accessed November 2007)

Richards, P. (1985) *Indigenous Agricultural Revolution: ecology and food crops in West Africa*, London: Hutchinson.

Robinson, D., Hewitt, T. and Harriss, J. (2000) *Managing Development: understanding inter-organizational relationships*, London: Sage.

Robson, C. (1993) *Real World Research: a resource for social scientists and practitioner-researchers*, Oxford: Blackwell.

Rodney, W. (1972) *How Europe Underdeveloped Africa*, London: Bogle L'Ouverture.

Rosenthal, U. (1998) 'Future Disasters, Future Definitions', in Quarantelli, E.L. (ed.) What Is a Disaster? Perspectives on the question, London: Routledge, pp. 146–159.

Rostow, W.W. (1960) *The Stages of Economic Growth: a non-communist manifesto*, Cambridge: Cambridge University Press.

Rubin, F. (1995) *A Basic Guide to Evaluation for Development*, Oxford: Oxfam.

Sachs, W. (1993) *Global Ecology: a new arena of political conflict*, London: Zed Books.

Sachs, W. (1999) *Planet Dialectics: explorations in environment and development*, London: Zed Books.

Saith, A. (2006) 'From Universal Values to Millennium Development Goals: lost in translation', *Development and Change* 37:6, pp. 1167–1199.

Scanlon, J. (1998) 'Dealing with Mass Death after a Community Catastrophe: handling bodies after the 1917 Halifax explosion', *Disaster Prevention and Management* 7:4, pp. 288–304.

Scheyvens, R. and Storey, D. (2003) *Development Fieldwork: a practical guide*, London: Sage.

Schipper, L. and Burton, I. (eds)(2008) *Earthscan Reader in Adaptation to Climate Change*, London: Earthscan.

Schipper, L. and Pelling, M. (2006) 'Disaster Risk, Climate Change and International Development: scope for, and challenges to, integration', *Disasters* 30:1, pp. 19–38.

Schön, D.A. (1983) *The Reflective Practitioner: how professionals think in action*, London: Temple Smith.

Schumacher, E.F. (1973) *Small Is Beautiful: economics as if people mattered*, London: Blond and Briggs Ltd.

Selman, P. (1996) *Local Sustainability: managing and planning ecologically sound places*, London: Paul Chapman.

Sen, A.K. (1981) *Poverty and Famines: an essay on entitlement and deprivation*. Oxford: Clarendon Press.

Shears, P. and Lusty, T. (1987), 'Communicable Disease Epidemiology Following Migration: studies from the African famine', *International Migration Review* 21:3, pp. 783–795.

Shiwaku, K. and Shaw, R. (2008) 'Proactive Co-learning: a new paradigm in disaster education', *Disaster Prevention and Management* 17:2, pp. 183–198.

Shiwaku, K., Shaw, R., Kandel, R.C., Shrestha, S.N. and Dixit, A.M. (2007) 'Future Perspective of School Disaster Education in Nepal', *Disaster Prevention and Management* 16:4, pp. 576–587.

Slovic, P. (2000) *The Perception of Risk*, London: Earthscan.

Smillie, I. and Minear, L. (2004) *The Charity of Nations: humanitarian action in a calculating world*, Bloomfield: Kumarian Press, Inc.

Smith, K. (2001) *Environmental Hazards: assessing risk and reducing disaster*, third edition, London: Routledge.

Sphere Project (2004) *Humanitarian Charter and Minimum Standards in Disaster Response,* Geneva: Sphere Project.

Smith, K.R. (2001) 'The Risk Transition and Developing Countries', in Kasperson, J.X. and Kasperson, R.E. (eds) *Global Environmental Risk*, London: Earthscan.

Smith, K.P. and Watking, S. (2004) 'Perceptions of Risk and Strategies for Prevention: response to HIV/AIDS in rural Malawi', *Social Science & Medicine* 60:3, pp. 649–660.

Smith, W. and Dowell, J. (2000) 'A Case Study of Co-ordinative Decision-making in Disaster Management', *Ergonomics* 43:8, pp. 1153–1166.

Stock, R. (1986) "Disease and Development" or 'the Underdevelopment of Health": a critical review of geographical perspectives on African health problems', *Social Science and Medicine* 23:7, pp. 689–700.

Strategy Unit (2002) *Risk: improving government's capability to handle risk and uncertainty*, London: UK Cabinet Office.

Summerfield, D. (2000) 'War and Mental Health: a brief overview', *British Medical Journal* 321: 232–235.

Summerfield, D. and Hume, F. (1993) 'War and Posttraumatic Stress Disorder: The question of social context', *Journal of Nervous and Mental Disease* 181:8, p. 522.

Summerfield, D. and Toser, L. (1991) '"Low intensity" War and Mental Trauma in Nicaragua: a study in a rural community', *Medicine and War* 7, pp. 84–99.

Tallis, H., Kareiva, P., Marrier, M. and Chang. A. (2008) 'An Ecosystem Services Framework to Support Both Practical Conservation and Economic Development', *Proceedings of the National Academy of Sciences* 105:28, pp. 9457–9464.

Tapsell, S. and Tunstall, S. (2006) 'The Mental Aspects of Floods: evidence from England and Wales, in Few, R and Matthies', F. (eds) *Flood Hazards and Health: responding to present and future risks*, London: Earthscan.

Thywissen, K. (2006) *Components of Risk: a comparative glossary*, Bonn: UNU-EHS.

Tjallingii, S.P. (1995) *Ecopolis Strategies for ecologically sound urban development,* Leiden: Backhuys.

Tobin, G.A. and Montz, B.E (1997) *Natural Hazards: explanation and integration*, New York and London: Guilford Press.

Todaro, M. (1981) *Economic Development in the Third World*, London: Longman.

Tomkins, A. and Watkins, F. (1989) '*Malnutrition and Infection*', ACC/SCN State-of-the Art Series, Nutrition Policy Discussion Paper 5, United Nations Administrative Committee on Coordination – Sub-Committee on Nutrition, October.

Toole, M.J. (1995) 'Mass Population Displacement. A global public health challenge', *Infectious Disease Clinics of North America* 9:2, pp. 353–366.

Toole, M.J. and Waldman, R.J. (1993) 'Refugees and Displaced Persons: war, hunger and public health', *Journal of the American Medical Association* 270:5, pp. 600–605. Toronto: University of Toronto Press.

Tudor Rose (2005) *Know Risk*, www.tudor-rose.co.uk: Tudor Rose.

Tudor Rose (2006) *Real Risk*, www.tudor-rose.co.uk: Tudor Rose.

Tudor Rose (2008) *Risk Wise*, www.tudor-rose.co.uk: Tudor Rose.

Twigg, J. (2007a) Disaster Reduction Terminology: a common sense approach, *Humanitarian Exchange,* 38, pp. 2–5.

Twigg, J. (2007b) 'Characteristics of a Disaster Resilient Community'. Version 1, for the DFID Disaster Risk Reduction Interagency Coordination Group, Available at http://www.benfieldhrc.org/disaster_studies/projects/communitydrrindicators/community_drr_indicators_index.htm.

UNDP (2002) *Human Development Report 2002*, Oxford: Oxford University Press.

UNISDR (2005) *Hyogo Framework for 2005–2015: building the resilience of nations and communities to disasters*, available at: www.unisdr.org/wcdr/intergover/official-doc/L-docs/Hyogo-framework-for-action-english.pdf (accessed 22 June 2005).

UNISDR (2007) *Terminology: basic terms of disaster risk reduction*, available at http://www.unisdr.org/eng/library/lib-terminology-eng%20home.htm.

United Nations (2008) *The Millennium Development Goals Report*, New York: UN.

United Nations (2003) *World Population Prospects: the 2002 revision*, New York: United Nations Press.

United Nations (2008) *The Millennium Development Goals Report*, New York: United Nations Press.

UNAIDS/WHO (2007) *AIDS Epidemic Update*, Geneva: UNAIDS and WHO.

UNDP (1994) *Human Development Report 1994*, Oxford: Oxford University Press.
UNDP (1995) *Human Development Report 1995*, Oxford: Oxford University Press.
UNDP (1997) *Human Development Report 1997*, Oxford: Oxford University Press.
UNDP (2003) *Human Development Report 2003*, Oxford: Oxford University Press.
UNDP (2004) *Reducing Disaster Risk: a challenge for development*, Geneva: Bureaux for Crisis Prevention and Recovery.
UNDP (2007) *Human Development Report*, Geneva: UNDP.
UNEP (2002) *Global Environment Outlook 3: past, present and future perspectives*, London: Earthscan.
UNHCR (2006) *Refugees by Numbers 2006*, Geneva: UNHCR.
UNICEF (1990) *Strategy for Improved Nutrition of Children and Women in Developing Countries. A UNICEF policy review*, New York, UNICEF.
UNICEF (2003) 'The UNICEF Nutrition Framework', the Sphere Project, 2004, Geneva: Sphere Project.
UNISDR (United Nations International Strategy for Disaster Reduction) (2004) *Living with Risk: A global review of disaster reduction initiatives,* Geneva: UNISDC.
Uphoff, N. (1993) 'Grassroots Organizations and NGOs in Rural Development: opportunities with diminishing states and expanding markets', *World Development* 21:4, pp. 607–622.
van Aalst, M.K. (2006) 'The Impacts of Climate Change on the Risk of natural disasters', *Disasters* 30:1, pp. 5–18.
Von Kotze, A. and Holloway, A. (1996) *Reducing Risk: participatory learning activities for disaster mitigation in Southern Africa*, Geneva ICRC.
Wadsworth, Y. (1997) *Do it Yourself Social Research*, second edition, St Leonards Australia: Allen and Unwin.
Walker, P. (1983) *Famine Early Warning Systems: victims and destitution*, London: Earthscan.
Wallerstein, I. (1979) *The Capitalist World Economy*, Cambridge: Cambridge University Press.
Wathern, P. (ed) (1988) *Environmental Impact Assessment: theory and practice,* London: Allen and Unwin.
Waugh, W.L. Jr and Streib, G. (2006) 'Collaboration and Leadership for Effective Emergency management', *Public Administration Review* 66:S, pp. 131–140.
WCDR (World Conference on Disaster Reduction) (2005) *Hyogo Accord*, Kobe, Japan: WCDR.
WCED (World Conference on Environment and Development.) (1987) *Our Common Future (The Brundtland Report)*, Geneva: WCED.
Weiss, R.A. and McMichael, A.J. (2004) 'Social and Environmental Risk Factors in the Emergence of Infectious Diseases', *Nature Medicine Supplement* 10:12, pp. 70–76.
White, P. (2000) 'Editorial: Complex Political Emergencies: grasping contexts, seizing opportunities', *Disasters* 24:4, pp. 288–290.
Whitehead, M. and Bird, P. (2006) 'Breaking the Poor Health-Poverty Link in the 21st Century: do health systems help or hinder?, *Annals of Tropical Medicine and Parasitology* 100:5/6, pp. 389–399.
Whiteside, A. and Sunter, C. (2000) *AIDS: the challenge for South Africa*, Cape Town: Human and Rousseau.
WHO (1948) *The Constitution*, Geneva: WHO.
WHO (1978) *Alma Ata 1978: primary health care*, Health for All series No.1, Geneva: WHO.
WHO (1991) *Community Involvement in Health Development: challenging health services, report of a WHO study group*, WHO Technical Report Series 809, Geneva: WHO.
WHO (1997) *World Health Report*, Geneva: WHO.
WHO (2003) *World Health Report,* Geneva: WHO.
WHO (2005a) *World Health Organization Statement on the Kyoto Protocol to the UN Framework Convention on Climate Change*, Geneva: WHO.
WHO (2005b) *World Health Report*, Geneva: WHO.

WHO (2005c) *Health and the Millennium Development Goals*, Geneva: WHO.

WHO (2006) *Epidemic and Pandemic Alert Response*, Geneva: WHO.

WHO (2007) 'A Safer Future: global public health security in the 21st century', *World Health Report*, Geneva: WHO.

WHO (2008a) 'Global and Regional Food Consumption Patterns and Trends', http://www.who.int/nutrition/topics/3_foodconsumption/en/index.html (accessed 30 July 2008).

WHO (2008b) 'Horn of Africa, Health Action in Crises', *Highlights* No 218, (21–27 July), Geneva: WHO.

Wikipedia (2008) Hurricane Katrina, synopsis at http://en.wikipedia.org/wiki/Hurricane_Katrina#cite_note-TPInteractive-3, (accessed September 2008).

Willems, R. (2005) 'Coping with displacement: social networking among urban refugees in an East African context', in Ohta, I. and Gebre, D. (eds) *Displacement Risks in Africa: refugees, resettlers and their host population*, Kyoto: Kyoto University Press and Trans Pacific Press, pp. 53–77.

Willis, K. (2005) *Theories and Practices of Development*, London: Routledge.

Wilson, G.A. and Bryant, R.L. (1997) *Environmental Management: new directions for the twenty-first century*, London: UCL Press.

Wilson, M.E. (1995) 'Infectious Diseases: an ecological perspective, *British Medical Journal* 311, pp. 1,681–1,684.

Winch, P.J., Makemba, A.M., Kamazima, S.R., Lwihula, G.K., Lubega, P., Minjas, J.N. and Shiff, C.J. (1994) 'Seasonal variation in the perceived risk of Malaria: implications for the promotion of insecticide-impregnated bed nets'. *Social Science and Medicine,* 39:1, pp. 63–75.

Wisner, B. and Adams, J. (eds) (2003) *Environmental health in emergencies and disasters: a practical guide*, Geneva: WHO.

Wisner, B., Blaikie, P., Cannon, T. and Davis, I. (2004) *At Risk: natural hazards, people's vulnerability and disasters*, second edition, London: Routledge.

Wood, A., Apthorpe, R. and Borton, J. (eds) (2001) *Evaluating International Humanitarian Action: reflections from practitioners*, London: Zed Books.

Young, H. and Jaspars, S. (1995) *Nutrition Matters: people, food and famine*, London: Intermediate Technology Publications.

Zetter, R. (2007) More Labels, Fewer Refugees: remaking the refugee label in an era of globalization', *Journal of Refugee Studies* 20:2, pp. 172–192.

Zschau, J. and Küppers, A.N. (eds) (2003) *Early Warning Systems for Natural Disaster Reduction*, Berlin: Springer.

Index

accidents and poisonings 125
accountability 40, 239–41
acid deposition 158
Action Aid 182
action research 179
activism 253; activist 260
adaptation 4, 59–61, 94, 104, 227–8, 231, 251, 254, 259; adaptation to crisis 14, 89
adaptive capacity 229, 251
advocacy 251
agricultural development 79–80
aid and relief 8
ALNAP (Active Learning Network for Accountability and Performance in Humanitarian Action 200–1
Amnesty International 37
animal diseases 127
antiretroviral drugs 5
assets 4–5, 77–8, 99–101, 228
Asylum Seeker 107
avian flu (or bird flu) 5, 134
awareness 8, 202, 253–5

bacteria 135–7
Bangladesh 1, 7, 32, 66, 89, 105, 213
baseline monitoring 198–9
baseline survey 215
behavioural change 253–4
biological susceptibility 261
biomedical approaches 6, 159
Bretton Woods Conference 63
Brundtland Commission 16
BSE 127, 133–4, 138–41
building design 32, 227; building regulations 8; resistant buildings 231, 233, 251; earthquake resistant building 228
business continuity 41, 106, 222, 229, 236

calories 156–7
capable people 5, 14
capability poverty 71–2, 77–8
capacity (or capability) 4, 8, 20–2, 71–2, 202, 212, 214, 218, 230; capacity building 39
carrying capacity 15, 59
cause and effect 191, 197
Centre for Research on the Epidemiology of Disaster (CRED) 11
chaos 25
chemical poisoning 157–8
Chernobyl nuclear plant 94
cholera 133, 141–3, 169; cholera risk assessment 215
Christian Aid 35, 38, 111
Civil Contingencies Act (UK) 85, 102, 119, 229, 234–9, 261
civil protection 235
civil society 36, 40–1
CJD 5, 127, 133–4, 138–41

class 83, 244
climate change 2, 29, 53, 60, 93–4, 228, 233, 254; climate change and health 143–51
coastal erosion 53
Cold War 65–7
command and control approach 26, 39, 164, 179, 236–7
common property environments 52, 57
common standards 227
communicable disease *see* infectious disease
communication 105, 208; of risk 206
community based approaches 2, 105, 244, 255, 257; community based development 83
community based disaster management 177–8, 181, 261
community based disaster preparedness 102
community based disaster risk reduction 39, 102
community based natural resource management 40, 57–8
community based organisations (CBOs) 189
community cooperation 33
complexity 25, 84–5, 164; complex development 4, 7, 25
complex emergencies 244; complex political emergency 5, 84–6, 208, 257
complex systems 7, 142, 148
communication 245, 252; communication technology 206
conflict 2, 4, 11–12, 14, 21, 33, 74, 85, 115–16, 245, 253; conflict risk reduction 210
conservation 40, 52–3, 56
contingency plan 106
cooperation 245
cooperative behaviour 51–2
coping 4–5, 8, 11, 13, 14, 60, 103–06, 120–21, 149, 158–61, 186, 202, 245, 250–1, 257; coping with catastrophe 260–2
cost benefit analysis (CBA) 194–5, 197–8
cost-effectiveness 117, 222, 261
critical infrastructure 2, 106
crop diseases 128
cultivation 23
culture 99, 102, 252, 259; cultural approaches 6, 26, 165; cultural construction of risk 218; cultural interpretation 261
cyclones 2, 89
Cyclone Nargis 2, 33, 119
Cyclone Preparedness Programme (CPP) 213
cyclone shelters 8, 89, 105–06, 213
Cyclone Sidr 1, 213, 244

dams 109, 195
debt 244
decentralisation 40
decision making 222
deforestation 2, 59
degenerative disease 127, 149
demographic aging 126
demographic transition model 125
Department for International Development (DFID) 11, 17, 76–7, 99
dependency 32, 64–70
determinism 23–5, 48
development: bottom up 39
development aid 33–5, 40, 67, 69
development goals 15
development indicators 87, 94
development induced disaster 28–32, 51–9, 195, 260
development institutions 1, 6
development out of disaster 244, 250
development paradigms 4, 28; development perspectives 8
development policy 34
development projects 20
development through disaster reduction 32–3, 120–21
diarrhoeal disease 5, 146
disaster and development 3–8, 14–15, 18–28, 32, 39, 42–7, 60, 86–7, 88–122, 148, 178, 207, 214, 217–8, 250–62
Disaster and Development Centre (DDC) xii, 180, 187–8, 191, 257
disaster and development industry 16, 32–5, 116
disaster and development studies xii, 3, 13, 203, 260
disaster avoidance 4, 7–8, 209, 221, 224, 251
disaster data 11, 90, 256
disaster definitions (or typology or classification or interpretations) 8, 11–13, 26
disaster education 7, 181–5, 262; people centred risk management education 202
Disasters Emergency Committee (DEC) 41, 200
disaster: environmental 21
disaster categorisation 90–2
disaster expenditure 8
disaster forensics 252

disaster impact 4–5, 8, 17–20, 88–122, 220–01, 231
disaster intervention 20–1, 188–9, 239
disaster management xii, 6, 8, 19, 28, 41, 86, 120–21, 177–8, 206, 222, 252, 256
disaster management cycle 6, 26–8, 169, 228
disaster mitigation 4, 6, 27–8, 31, 89, 178–9, 209, 214, 217, 227–49
disaster preparedness 7, 27, 30, 207, 211–12, 217, 224, 256–7
disaster prevention 1, 7–8, 15, 19, 27, 40, 42, 207, 248, 251, 255
disaster recovery 2, 4–6, 7–8, 27–8, 31, 227–49
disaster reduction 1, 2, 6–8, 14–15, 27, 32, 224, 251, 262
disaster reporting 1, 8, 12
disaster research 176–84, 202, 254–6, 259, 261–2
disaster resilient communities 230–4, 256–7, 259
disaster resistant infrastructure 231–4
disaster response 1, 5–8, 27–8, 222, 227–49; strategies 209
disaster risk 20, 23, 253
disaster risk reduction (DRR) 6, 8, 17, 41–2, 144, 182, 205, 207, 216, 218, 221, 224, 230, 254–5, 259–60, 262; disaster risk reduction = sustainable development 218, 250; disaster risk reduction and governance 219–33; learning in DRR 202
disaster sociology 26
disaster trends 8. 11
disease ecology 138
disease hazards 116–17, 125
disease risk 5
disease transmission 125
displacement: development induced 2; mass displacement 30
drought 2, 6

early warning 6–7, 27, 30, 118, 180, 205–26, 252, 262; of disease epidemic 209–10
earthquake 2, 7, 13, 32, 95, 119; Pakistan earthquake 228
economic change 23
economic collapse 16, 61–2
economic development 40, 61–4, 82, 96–7, 125
economic impact of disasters 33
ecological perspective 48, 53
ecosystems 25, 40, 136
education 17 *see also* disaster education
EM-DAT 11, 90
emergency management 28, 248
emergency preparedness 197, 221–2
emergency relief 5, 7, 69, 117, 227–9, 245, 262
emergency responders 234; coping of emergency responders 234
emergency response 211, 227; integrated emergency response 227
emergency services 36, 85, 103, 106, 179, 229–30
emergent and resurgent disease 133, 138
empowerment 58, 99, 181–3, 200
endemic disease 5, 130
El Nino Southern Oscillation (ENSO) 144–5
entitlements 50, 85, 99, 154, 208; entitlement mapping 216
environment 13, 71
environmental assessments 215
environmental catastrophe 5
environmental change 14, 23, 210
environmental degradation 4, 14–16, 50–3, 56, 59, 71–5
environmental disasters 92
environmental footprint 251
environmental hazards 2, 13, 21, 212, 251
environmental health 116, 169–70
environmental impact assessment (EIA) 192–4, 197
environmental management 22, 158, 219
environmental security 99
environmental systems 25, 94, 198
environmentalism 50
epidemics 5, 130, 134, 140–2, 209–10
epidemiology 125, 135–6, 140
epidemiological transition 125
ethics 41, 60, 239
ethnic cleansing 115
ethnography 184
European Union 34
evaluation 6, 199–200, 230, 259
evolution 24
expert knowledge 2–3

faith 38, 252, 262
famine 6, 17, 21, 49–50, 70, 115, 153–7, 208
famine early warning systems (FEWS) 7, 208, 215
Federal Emergency Management Bureau (FEMA) 26, 30–1, 85, 119, 235
floods 11, 92–3, 232–3; flood protection 32
Food and Agriculture Organisation (FAO) 49, 215
food security 6, 17, 49–50, 99, 102, 148, 153–7, 229; food security information system 208

food supply 49–50, 156–7, 215
foot and mouth disease 139
forced displacement 106–16
Fred Cuny 229

gatekeepers 20
gender and development 73–4
gender and disaster 82, 230, 246
Gender Empowerment Measure (GEM) 74
gender inequality 153
genocide 108–9
geographic information systems (GIS) 196–7, 215–6
geography 2
geophysical hazards 91
global change 23
global warming 29, 144
globalisation 4, 23, 55, 78–84, 127, 245
gold, silver, bronze command 229, 337
governance 7–8, 14, 21, 41, 57–8, 70, 118, 137, 205, 224, 246, 250; good governance 248, 258–9; local governance 181–2, 187
Great Depression of the 1930s 62

hazards 214, 217–8, 224, 231, 257; natural hazards 13, 19
health 2, 5, 12, 17, 123–75, 261; health and sustainable development 128–29, 164
health belief model 165–6
health care 161–4; emergency health care 167–71; environmental health care 169; preventative health care 6, 130; primary health care (PHC) 6, 162–3, 167–71; self care 161–4; traditional health care 161
health disaster 20
health ecology 136–8, 171
health hazards 130–31, 135–8
health impacts 115–17; impact of disaster on health services 170
health interventions 138, 148, 165
health prevention 124
health promotion 164–7
health risks 130, 145, 158; health risk reduction 172–3
health security 99, 102, 136, 150–1, 232, 251
hill slopes 8
HIV/AIDS 5, 11, 61–2, 95, 133, 151–3, 166–7
H5N1 127, 138–41
Homeland Security 85, 235
Horn of Africa 85, 111
human agency 25–6, 130, 167
human behaviour 2, 15, 25, 52–3, 115, 137, 251, 253
Human Development Index (HDI) 21, 95–7
Human Poverty Index 21
human security 98–103
human settlement 8
human systems 7
humanitarian aid and assistance 11, 21, 33–5, 42, 69–70, 101, 115–20, 154, 200, 229, 239–41, 245
humanitarianism 7, 85, 227, 230, 239–41
Humanitarian Charter 230, 240
hurricanes 7, 211–12; Hurricane George 244; Hurricane Katrina 30–1, 85, 95, 211–12; Hurricane Mitch 70, 211, 244; Hurricane Michelle 211; Hurricane Ivan 211
Hyogo Framework for Action 40, 42–3, 103, 187–8, 230, 261

impact assessment 6, 178, 192–9; impact assessment methodology 196–7
implementation 190, 195, 202, 207, 259
indicators 7, 16, 20–1, 134, 190–1, 198, 208–9; SMART indicators 201
indigenous knowledge 207
industrialisation 29, 50, 251
infectious disease 2, 5–6, 11–12, 116, 124–5, 130–38; infectious disease complexity 135–6; infectious disease risk management / reduction (IDRM / IDRR) 133–38, 143, 188
info-gap decision theory 217
information systems 252, 256
information technology 215, 230
infrastructure (resistant) 7
innovation 60
insecurity 261
insurance 95, 222–3; insurance industry 222
institutional strengthening 8
integrated approaches 2–3, 8, 48, 110, 124, 135–7, 143, 151, 171, 259; research 180–7
integrated emergency management (IEM) 229, 234–9
internally displaced people (IDPs) 108–11
international accords 15
International Federation of the Red Cross and Red Crescent Societies (IFRC) 11, 91, 185–6, 211
International Monetary Fund (IMF) 34, 41, 70, 244

International Panel on Climate Change (IPCC) 94, 104, 110, 145, 150
international organisations (IOs) 11, 20, 34, 188, 190
investment 8, 259
Iraq 114

Japan 7, 32

knowledge, attitudes and practice (KAP) 166
Kyoto Protocol 145, 254–5

land degradation 17
learning by doing 6–7, 176–7
learning cycles for disaster and development 177–81
legislation 195, 202, 223, 254
lessons learnt 6, 27, 179, 239, 252
limits to growth 59–61
livelihoods 4, 23, 58, 77–8, 233, 246; livelihood security 148, 155, 209; livelihood system 23
local development 16, 57
local knowledge 2–3, 179–85, 188, 259; local solutions 22, 40
logical framework analysis 190–1

malnutrition 133, 148, 153, 156–7
Malthusian approaches 4, 15, 48–50, 55, 62, 124–5
mangroves 8
marginalisation 28, 73, 99
market approaches 61–2
Marxist ideas 26, 63–9
mass emergency 209, 230
media 11, 118–19
Médicins Sans Frontières (MSF) 35, 37, 170, 209
mental health 5, 115, 124, 158–61; mental impairment 261
meteorological crisis 211
micro-credit 244
Millennium Development Goals (MDGs) 16–17, 20, 32, 41, 118, 153, 163, 188, 251
minimum standards 7, 194, 230, 239–43
modelling 197
moral imperative 221; moral issues 239, 244
multileveled approach 22–3, 245
Monitoring and Evaluation 6, 40, 42, 90, 178–9, 195, 199–202; participatory monitoring and evaluation 200
morbidity 5
mortality 5
motivation 207, 259
Mozambique 10, 41, 68–9, 79–80, 113; Mozambique floods 31, 33, 70, 97–8, 145, 170, 231–3; and disease 141–3
multidisciplinary 259

national development plans 16, 41–2
'natural' disaster 1, 11, 13–15, 91
natural hazards approach 47–8; natural hazards research 180; natural systems 25
needs 6, 14, 71, 115; needs assessment 178–9, 185–8
Nepal 7, 33
non-governmental organisations (NGOs) 20, 35–41
nutrition 6, 49, 146, 148, 153–7

obesity 127, 153–4, 254
offsetting 56
origins of disaster 14–15, 32, 35
Organisation for Economic Cooperation and Development (OECD) 33, 71–2, 201
Oxfam 35, 37, 157
ozone layer 158

participation 24, 39, 143, 163, 177–85, 244, 259; participatory action and research 255; participatory appraisal 179, 183, 203; participatory approaches 252; participatory evaluation 199
pathogens 125, 135–8
pathways 136, 143, 219
people centred hazard and vulnerability mitigation/reduction 180, 188, 257–8, 259
perception 2, 7, 23, 26, 136–8, 165–7; perceived risks 103
personal development 23
pesticides 128
phenomenology 24
planning 6, 176–204
policy environment / regimes 5, 118–20, 188, 190, 202
political divisions 21
political ecologies 138, 147
political ideology 30
political or personal will 2, 7, 206, 212, 221, 251, 253–4
pollution 7, 29, 50, 55–6, 93, 147, 158, 219, 251
population 11, 15, 49–50, 55, 62, 86, 125–6;

population density 25; population displacement 106–15, 232–3
positivism 24
possibilism 24
post conflict 246
post disaster 6
post-structuralism 26
poverty 4–5, 17, 25, 62, 71–5, 253, 258; poverty alleviation 75–8; poverty and health 137–8; poverty assessment 221; poverty cycle 71–5, 148–9; poverty reduction 14–15, 20, 71–82, 221, 253
Poverty Reduction Strategy Papers (PRSPs) 41, 75
practice led 6
precautionary principle 7, 144, 149, 219, 254
prediction 7, 180, 206, 208, 222, 224; predictive models 7, 205, 210, 214–5, 252; for earthquake 214
pre disaster 28
prevention strategies 2, 118, 162, 167; preventive medicine 129
probability 217
project cycle 6, 178–9
project planning 190–1
protection 11, 228–9, 231
psychosocial aspects of disaster 6, 92, 99, 110, 115, 129, 146, 158–61, 253
push and pull factors 232

quantitative surveys and techniques 179–81, 215, 217

rapid onset disaster 1, 7, 16–17, 89, 228
reconstruction 228–9, 246; reconstruction balance sheet 244–7
recovery 118
recycling 158
reductionist approach 25
reflection in action 177–9
refugees 5, 82, 106–15
refugee cycle 112–14
rehabilitation 27
relief 2, 27–8, 32, 92, 116–18, 154; relief agencies 200; relief and development 251
religious belief 38, 262
remote sensing 30; remotely sensed data 209
representation 5, 99, 117–19, 144, 183, 248, 252, 255, 258–9
resettlement 5, 115–6
resilience 5, 8, 14–15, 102–106, 120–21, 144, 148, 179, 202, 187, 218–19, 221, 224, 227–9, 231, 234, 245, 251–2, 256–7, 261–2
resilience capacity assessment (RCA) 187
Resilience Forums (UK) 40, 236
resilience through wellbeing 259, 261
Resilience UK 234
resilient communities 40, 116, 213
resistant buildings 177
responders 229–30; Category 1 and 2 responders 235–9
responsibilities 250, 253
restoration 33, 227–8
review 6
rights 2, 8, 17, 21–22, 50, 99, 102, 107, 144, 230, 248, 250–52, 254–56, 258–9
risk 4, 199, 216–19, 223–4; theoretical risk 139; reaction to risk 206, 223; risk thresholds 206; risk perception 206–7, 222; risk and resilience 151, 187, 258–9
risk and resilience committees (RRC) 106, 143, 187–91, 245, 257–8
risk assessment 6–7, 30, 42, 182, 189, 199, 206, 214, 219, 221–3
risk communication 182, 187–9
risk governance 4, 8, 42, 105, 222–3, 251, 258, 262
risk management 6–7, 141, 205–26, 222–4
risk mapping 215–6
risk mitigation 232–3
risk reduction 85, 105, 120, 206, 224, 250, 255–6; see also disaster risk reduction
risk transfer 223
risky behaviour 56

safety 40, 229–31
SARS 5, 127, 134, 138–41
Save the Children 38
scenario analysis 197
science and technology 251–4
seasonality 145, 210
sea level rise 233
search and rescue 252
security 246, 251, 253, 260–2: human security 4, 8, 48, 260–2; international security 30
self-reliance 167
simulation exercises 179–80
Sit Reps (situation reports) 179
slow onset disaster 6, 16–17, 89, 228
social care 161–4

social change 23
social constructionism 26
social decay 16
socio-economic wellbeing 137
social impact assessment (SIA) 192, 194
social justice 37
social structures 23
social systems 198
Sphere guidelines 7, 42, 129, 157, 171, 230, 239–43
Sri Lanka 21–2, 70, 120
storms 11
Structural Adjustment Programmes (SAPs) 67–9
structural approaches 25, 48, 130, 165
sub-Saharan Africa 5, 62, 70, 131–32, 152
Sudan 33
surveillance 7, 214–15; anthropometric surveillance 215
survival 8, 13, 59, 94, 221; survival strategies 2–3
survivors 5, 115–16, 159, 179, 245
susceptibility 8, 148
sustainable development 1, 4, 8, 14–16, 27–8, 41, 47–8, 77, 86–7, 120–21, 177–8, 201, 218, 229, 248, 254–6, 260, 262
sustainable livelihoods 99–100, 114
systems approach 22–24, 197

table top (desk top) exercises 180, 197
technology 262; technological development 28, 30, 55; technological disasters 12, 14
tectonic activity 13
telecomunications 2
terrorism 85
thresholds 194, 208
topocide 109
torture 115
tourism 92, 102
tradition and belief 7
trade 56, 62–3, 155, 208
Tragedy of the Commons 50–9, 86
trauma 6, 115, 158–61, 232; post traumatic stress 159
Truth and Reconciliation Commission 70
tsunami 7–8, 21–2, 70, 89, 206

UK floods 95
UK resilience 102, 106
uncertainty 2, 7, 25, 134, 138–42, 148, 164, 205–6, 214, 216–21, 252, 256
underdevelopment 6, 16, 26, 28, 30, 60, 106, 251, 253–4
uneven development 6, 26, 30, 70, 134, 161
United Nations Children's Fund (UNICEF) 154–5, 182
United Nations Development Programme (UNDP) 11, 20, 39, 95, 98–9, 182
United Nations High Commission for Refugees (UNHCR) 107–08
United Nations International Strategy for Disaster Reduction (UNISDR) 11, 39, 42, 182, 233
United Nations Office for the Coordination of Humanitarian Affairs (OCHA) 98
urban disaster 81–4
urban poverty 80–3
urbanisation 4, 78–84, 127

values 24
VCA 6, 178, 185–8
volcanoes 94
vulnerability 4, 14, 15, 19, 28, 64–70, 87, 179, 202, 210, 214–18, 224, 231, 250–2, 260–2, 261; vulnerability of cities 82; vulnerability to disease 135–6, 146, 148; vulnerable groups 246; vulnerability of women 82–3; vulnerability reduction 6, 14

warfare 85
water quality 195
water resources management 24
wealth 82, 99
wellbeing 3, 8, 15, 30, 40, 46, 75, 99, 102–06, 124, 159, 186, 219, 231, 245, 250–1, 257, 260–2, 261–2
Wenchuan Earthquake 2, 33, 119
wildlife 58–59
willingness to pay 197
World Bank 34, 41, 70, 95, 244
World Conference on Disaster Reduction (WCDR) 42, 254
World Conference on Environment and Development (WCED) 16, 47, 254
World Conference on Sustainable Development (WCSD) 16, 254
World Development Report 20
World Disasters Report 11, 104
World Health Organisation (WHO) 2, 124, 129, 162
World Trade Organisation (WTO) 56

Yokohama Initiative 42